T5-CVT-174

Status and Management of Interior Stocks of Cutthroat Trout

Cover: Pen and ink drawing and cover design by Ellen Meloy, Helena, Montana.

Funding for publication of this book was provided by the

U.S. Bureau of Land Management

U.S. Fish and Wildlife Service

U.S. Forest Service

U.S. National Park Service

Status and Management of Interior Stocks of Cutthroat Trout

Edited by

Robert E. Gresswell

American Fisheries Society Symposium 4

Bethesda, Maryland
1988

The American Fisheries Society Symposium series is a registered serial. Suggested citation formats follow.

Entire book

Gresswell, R. E. 1988. Status and management of interior stocks of cutthroat trout. American Fisheries Society Symposium 4.

Article within the book

Gerstung, E. R. 1988. Status, life history, and management of the Lahontan cutthroat trout. American Fisheries Society Symposium 4:93–106.

Library of Congress Catalog Card Number: 88–72021
ISBN 0-913235–55–5 (cloth) ISSN 0892–2284
ISBN 0-913235–56–3 (paper)

Address orders to

American Fisheries Society
5410 Grosvenor Lane, Suite 110
Bethesda, Maryland 20814, USA

CONTENTS

Preface

This book is the result of many years of work by numerous individuals. The idea for producing a synopsis of information concerning interior stocks of cutthroat trout originated from a special session on the subject at the 1979 Annual Meeting of the American Fisheries Society in West Yellowstone, Montana. John S. Griffith assumed the responsibility for organizing a publication from that session.

After several years, it became apparent that the initial effort had become stalled, and Griffith organized a session highlighting interior cutthroat trout at the 1986 meeting of the Western Division of the American Fisheries Society in Portland, Oregon. At that time, all of the original contributors were invited to submit updated manuscripts, and additional papers, reflecting a broadening of the fundamental cutthroat trout data base, were solicited. The session resulted in a preliminary compilation of available information; however, several of the less-abundant stocks were not included.

At the 1987 Western Division meeting in Salt Lake City, Utah, Bryce Nielson organized a session that included the most recent information concerning the less-prominent stocks of cutthroat trout. Papers from that session were added to those from the previous year, and the current volume emerged.

Donald Martin, President of the Western Division in 1986–1987, provided the impetus to complete the project in a timely manner. Martin made the initial contacts necessary to organize the project, and he continued to work with Alvin Mills, who succeeded him as president, to secure the funding necessary for publication.

This volume summarizes current information concerning the status and management of interior stocks of cutthroat trout. Although we have not included all subgroups, the major subspecies and most of the minor ones are discussed. Additionally, papers providing a brief history of the discovery of cutthroat trout, an overview of the taxonomic status of the species, and a general discussion of the interactions of cutthroat trout with other salmonids are presented.

ROBERT E. GRESSWELL, *Technical Editor*
U.S. Fish and Wildlife Service
Post Office Box 184
Yellowstone National Park,
Wyoming 82190, USA

Editorial Acknowledgments

Each manuscript was criticized by at least three peer reviewers, the symposium editor, and the American Fisheries Society's managing editor. The following reviewers contributed substantially to the quality of this publication.

Gaylord R. Alexander
Reeve M. Baily
Eric P. Bergersen
N. Allen Binns
Peter A. Bisson
Daniel G. Carty
Theodore M. Cavender
Patrick D. Coffin
Kurt D. Fausch
John J. Fraley
William R. Gould
Richard W. Gregory
William T. Helm
Thom H. Johnson
Calvin M. Kaya
David L. Langlois
Robb J. Leary
Eric J. Loudenslager
Virgil K. Moore
Robert F. Raleigh
Bradley B. Shepard
Gerald R. Smith
Russell F. Thurow
Ray J. White
Robert W. Wiley

Nomenclatural Changes

As noted by R. J. Behnke on pages 1–2 of this volume, the validity of *Salmo* as a generic name for several western North American trout species has been questioned in recent years. Taxonomists now agree that native "*Salmo*" trouts of northern Pacific Ocean drainages are more closely allied with Pacific salmon *Oncorhynchus* spp. than with Atlantic and Eurasian *Salmo* species (among which are Atlantic salmon *S. salar* and brown trout *S. trutta*). Recent evidence, culminating in new data presented during the June 1988 meeting of the American Society of Ichthyologists and Herpetologists, has persuaded the American Fisheries Society's Committee on Names of Fishes to accept *Oncorhynchus* as the appropriate generic name for all native Pacific-drainage trouts that presently are called *Salmo*. The evidence indicates that species of *Rhabdofario* (fossil trouts), *Parasalmo* (proposed to replace *Salmo* for living Pacific-drainage trouts), and *Oncorhynchus* are not distinctive at the generic level. Of these names, *Oncorhynchus* has historical taxomonic priority for this group of fishes.

A separate problem has concerned the specific name for rainbow trout (and its anadromous form, steelhead), which presently is called *Salmo gairdneri*. Taxonomists now believe that rainbow trout and the "Kamchatkan" trout *Salmo mykiss* of Asia form a single species, for which *mykiss* has nomenclatural priority. The Names of Fishes Committee thus has adopted *Oncorhynchus mykiss* as the scientific name of this species.

The North American species affected by these changes are as follows: Apache trout becomes *Oncorhynchus apache*, cutthroat trout becomes *O. clarki*, Gila trout becomes *O. gilae*, golden trout becomes *O. aguabonita*, Mexican golden trout becomes *O. chrysogaster*, rainbow trout becomes *O. mykiss*. The old names are used in the present volume, which was prepared before these questions were resolved, but the American Fisheries Society will implement the new names in its 1989 publications. Common names will remain unchanged.

ROBERT L. KENDALL, *Managing Editor*
American Fisheries Society

American Fisheries Society Symposium 4:1–7, 1988

Phylogeny and Classification of Cutthroat Trout

ROBERT J. BEHNKE

Department of Fishery and Wildlife Biology, Colorado State University
Fort Collins, Colorado 80523, USA

Abstract.—The cutthroat trout *Salmo clarki* and rainbow trout *S. gairdneri* and allied forms share a much closer phylogenetic relationship to Pacific salmons of the genus *Oncorhynchus* than they do to other species of the genus *Salmo*. Until the correct generic classification is decided, however, western North American trouts should be retained in *Salmo*. The cutthroat trout species initiated divergence into distinct evolutionary lines perhaps about 1 million years ago. Four "major" subspecies, *clarki* (coastal), *lewisi* (westslope), *bouvieri* (Yellowstone), and *henshawi* (Lahontan), represent ancient divergences and exhibit considerable genetic differentiation. Four "minor" subspecies have been derived from *S. c. henshawi*, and six minor subspecies have evolved from *bouvieri*, in relatively recent geological times (probably during the last glacial period). These minor subspecies exhibit diagnostic morphological differences, but show little or no quantifiable genetic divergence from the parental subspecies as determined by electrophoresis. The practical aspect of recognizing subspecific geographic races of cutthroat trout concerns the protection and preservation of biodiversity.

The cutthroat trout *Salmo clarki* has attained the broadest distribution of any species of trout (*Salmo*) in North America. It is the only trout native to the waters of Colorado, Wyoming, Utah, and Alberta, and it was the dominant native trout, in terms of distribution, in Nevada, Idaho, and Montana. The cutthroat trout was the first trout recorded by Europeans in North America. Coronado's expedition encountered an abundance of the Rio Grande subspecies in the Pecos River, New Mexico, in 1541 (Trotter 1987). The first trout encountered by the Lewis and Clark expedition in 1805 was a cutthroat from the falls of the Missouri River, Montana, a subspecies now recognized as *Salmo clarki lewisi*.

In the 19th century, as North American frontiers moved westward, the cutthroat trout became an important item of commerce and sustenance for the new settlers, as it had been for the native Americans for thousands of years. The impact of European civilization on cutthroat trout was rapid and devastating. In less than 100 years after the first settlements in the West, the cutthroat trout vanished from most of its vast range, except for the coastal subspecies, to be replaced by rainbow trout *Salmo gairdneri*, brown trout *S. trutta*, and brook trout *Salvelinus fontinalis* (and by lake trout *Salvelinus namaycush* and kokanee *Oncorhychus nerka* in many large lakes). State and federal resource agencies' historical attitudes and management policies in relation to cutthroat trout have ranged from benign neglect to outright extermination (Behnke 1981). Fortunately, this attitude is rapidly changing, as evidenced by the recent surge of interest in, and concern for, cutthroat trout by both professional biologists and laymen. Two recent books, written mainly for anglers (Smith 1984; Trotter 1987), extol the wonders and virtues of rare and beautiful native trout. Aside from our stewardship obligation to preserve native life, I hope this present volume propagates the message that the biodiversity contained among the 14 subspecies of cutthroat trout has significant potential for application in present and future management programs.

Nomenclature

The generic classification of rainbow and cutthroat trouts is presently in a state of indecision. The reason for this concerns the phylogenetic position of rainbow and cutthroat trouts in the subfamily Salmoninae. I have previously produced phylogenetic diagrams of Salmoninae showing closer relationships of the rainbow and cutthroat trout species to Pacific salmons of the genus *Oncorhynchus* than to brown trout or to Atlantic salmon *Salmo salar*, the type species for *Salmo* (Behnke 1968; Kendall and Behnke 1984). Electrophoretic analysis of proteins (Ferguson and Fleming 1983) and analysis of mitochondrial DNA base sequences with restriction endonucleases (Gyllensten and Wilson 1987) support the morphological–anatomical evidence for these alignments. There is no longer any reasonable doubt that rainbow trout and cutthroat trout are not phylogentic members of the genus *Salmo*.

With the recent strong emphasis in taxonomy to make classification reflect phylogeny, there is need for revision of the generic classification of cutthroat and rainbow trouts. In response to this, the genus *Parasalmo* was used by Kendall and Behnke (1984) for the evolutionary lineage of rainbow and cutthroat trouts. The name *Parasalmo* was first proposed by Vladykov (1963). Fossil trout and the rule of priority in taxonomic nomenclature, however, present a problem concerning the use of *Parasalmo*. A western North American fossil trout was described as a new genus, *Rhabdofario*, by E. D. Cope in 1870. As more North American fossil trout material has been critically studied, the former distinctions between *Rhabdofario* and *Parasalmo* have become blurred. In the future, *Rhabdofario* and *Parasalmo* are not likely to be recognized as separate genera and perhaps not even as separate subgenera (Gerald Smith, University of Michigan, personal communication). *Rhabdofario* clearly has priority over *Parasalmo* if both are considered synonymous.

Resolution of the uncertainty of correct nomenclature awaits the outcome of research on trout fossils. The American Fisheries Society's Names of Fishes Committee urges that until the relationships among *Oncorhynchus*, *Parasalmo*, and *Rhabdofario* are more clearly understood, rainbow and cutthroat trouts should remain in the genus *Salmo* (personal communications from Carl Bond, Robert R. Miller, and Gerald Smith).

Origins and Distribution

Recent studies on fossil trouts (Smith 1981; Taylor and Smith 1981; Cavender and Miller 1982; Smith and Miller 1985; Cavender 1986) indicate that separation of a "protosalmo" ancestor into a Pacific Ocean group and an Atlantic Ocean group occurred by the mid-Miocene, about 15 million years ago. The Atlantic Ocean group was the ancestor of the genus *Salmo* in the strict sense (i.e., Atlantic salmon and brown trout). The Pacific Ocean group divided into three evolutionary lines, *Oncorhynchus*, "*Parasalmo*," and *Rhabdofario*, by the end of the Miocene or about 5 million years ago. During the Pliocene, large-scale extinctions of freshwater fishes occurred in western North America and species of *Rhabdofario* and inland species of *Oncorhynchus* became extinct, to be replaced by species of "*Parasalmo*." Most of the early evolutionary lines of "*Parasalmo*" also became extinct, but one surviving line (the common ancestor) gave rise to the present species of cutthroat and rainbow trouts (as well as the golden and Mexican golden trouts *Salmo aguabonita* and *S. chrysogaster*, Gila trout *S. gilae*, and Apache trout *S. apache*). Sound evidence is lacking on the time when the common ancestor split into two lines, one leading to cutthroat trout and the other to rainbow trout and its allied forms. It can be speculated, however, based on electrophoretic and DNA evidence (Loudenslager and Gall 1980; Wilson et al. 1985; Loudenslager et al. 1986; Gyllensten and Wilson 1987; Leary et al. 1987; Leary and Allendorf 1988), that this split occurred near the end of the Pliocene or beginning of the Pleistocene, roughly 2 million years ago.

My interpretation of the available evidence indicates that the early major evolutionary divergences of cutthroat trout into four major subspecies groups were associated with the Columbia River basin. Evidently, the first two major divergences resulted in three characteristic karyotypes: 2N (diploid chromosome number) = 68, coastal cutthroat trout *S. clarki clarki*; 2N = 66, westslope cutthroat trout *S. c. lewisi*; and 2N = 64, ancestral Yellowstone cutthroat trout *S. c. bouvieri* (Figure 1). An early member of the 64-chromosome group gained access to the Lahontan Basin (in present-day Nevada) to give rise to the fourth major subspecies group (Lahontan cutthroat trout *S. c. henshawi*). I speculate that the separation of the three major subspecies groups, and perhaps the fourth (Lahontan Basin), occurred by the mid-Pleistocene or perhaps 1 million years ago.

Of the 14 subspecies of cutthroat trout I recognize, the 4 defined above can be considered a "major" subspecies in relation to the magnitude of genetic divergence and time since their evolutionary separation. The coastal and westslope cutthroat trouts did not give rise to any other surviving subspecies. In the Lahontan Basin, the ancestral cutthroat trout gave rise to the Paiute trout *S. c. seleniris* (a spotless form of Lahontan cutthroat trout isolated in Silver King Creek, California) in postglacial times. Three other undescribed subspecies were also derived from the Lahontan cutthroat: a fluviatile form indigenous to the Humboldt River drainage of the Lahontan Basin, and two subspecies established from transfers from the Lahontan Basin into the Alvord and Whitehorse basins (contiguous basins on the northern rim of the Lahontan Basin in northern Nevada and southern Oregon). These latter three subspecies probably originated during the early to

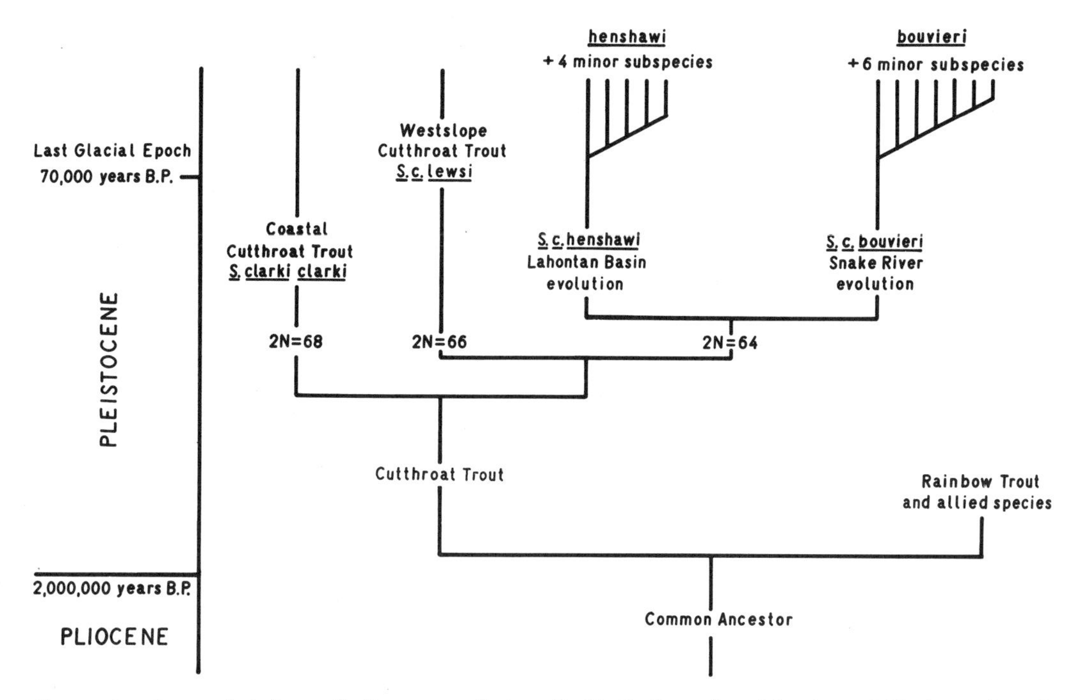

FIGURE 1.—Assumed phylogenetic divergences that resulted in the formation of 4 major and 10 minor subspecies of cutthroat trout. The diploid chromosome numbers (2N) of 68, 66, and 64 denote the first two divergences: first, a cutthroat trout ancestor split into the coastal subspecies, *S. c. clarki*, and the inland Columbia River group; then, the Columbia River group split into an upper basin subspecies, *S. c. lewisi*, and a 64-chromosome ancestor of the Snake River subspecies, *S. c. bouvieri*. The fourth major subspecies, *S. c. henshawi*, originated from an ancient penetration and subsequent isolation of the 64-chromosome ancestor in the Lahontan Basin.

middle period of the last glacial epoch, 40,000–70,000 years ago.

All other subspecies that I recognize were derived from a Yellowstone ancestor, probably in relatively recent times during the last glacial period. Interbasin transfers moved cutthroat trout from the Snake River into the Bonneville basin, giving rise to the Bonneville cutthroat trout *S. c. utah*, and into the Colorado–Green River basin, giving rise to the Colorado River cutthroat trout *S. c. pleuriticus*. Fish from the Colorado River basin crossed the Continental Divide and established native cutthroat trout populations in the headwaters of the South Platte and Arkansas rivers; these gave rise to the greenback cutthroat trout *S. c. stomias* and the now extinct yellowfin trout *S. c. macdonaldi* of Twin Lakes, Colorado. Another transfer into the Rio Grande basin gave rise to the Rio Grande cutthroat trout *S. c. virginalis*. An undescribed subspecies, the fine-spotted Snake River cutthroat trout, is indigenous to the upper Snake River, Wyoming. The fine-spotted form probably originated from a Yellowstone cutthroat trout ancestor isolated in an ice-dam lake during the late phase of the last glacial period (Malde 1965; Murphy 1974). Virtually no genetic differentiation can be detected by electrophoresis among the fine-spotted Snake River, the greenback, the Colorado River, and Yellowstone subspecies of cutthroat trout (Leary et al. 1987), which suggests that these forms diverged recently.

Crossings of the Continental Divide by westslope cutthroat trout into the South Saskatchewan drainage (Hudson Bay) and the upper Missouri River drainage and by Yellowstone cutthroat trout from the headwaters of the Snake River drainage into the Yellowstone River drainage occurred in postglacial times, as evidenced by the glacial geology of the areas (Roscoe 1974) and by the lack of differentiation between populations of the two subspecies on each side of the Continental Divide.

This brief account of cutthroat trout evolution and distribution is, admittedly, an oversimplification. How much is yet to be learned on the matter can be surmised from Rogers et al (1985), who discussed fossils of "cutthroat trout" from

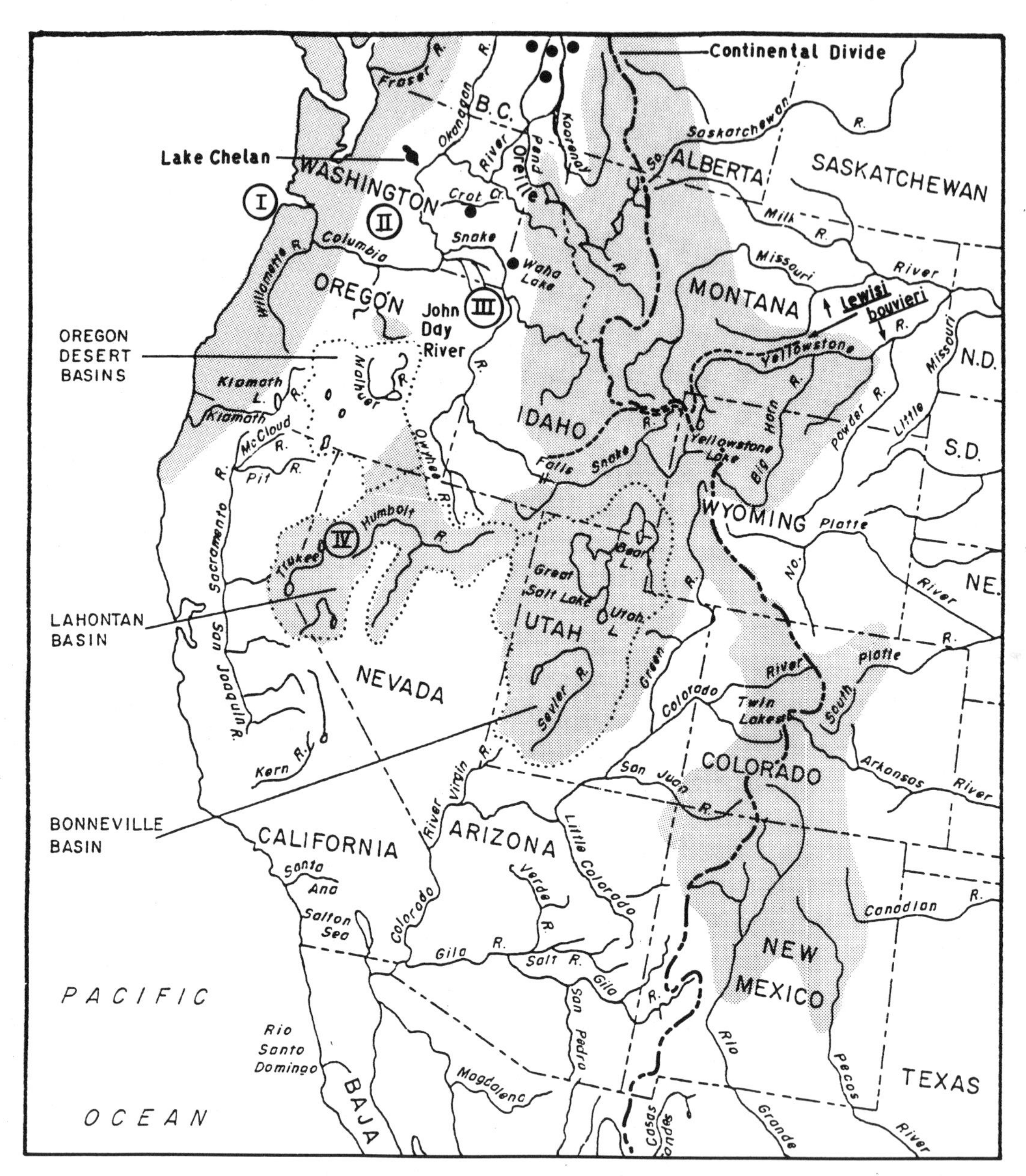

FIGURE 2.—Original distribution (shaded areas) of *Salmo clarki* (the northern distribution of coastal cutthroat trout extends off the map). Roman numerals indicate assumed areas of origins of the four major subspecies: **I**, coastal cutthroat trout *S. c. clarki*; **II**, westslope cutthroat trout *S. c. lewisi*; **III**, Yellowstone cutthroat trout *S. c. bouvieri*; **IV**, Lahontan cutthroat trout *S. c. henshawi*. Disjunct distributions of *S. c. lewisi* occur in the upper Columbia River basin (British Columbia, black dots), Lake Chelan (Washington), and John Day River (Oregon). Disjunct distributions of *S. c. bounvieri* occurred in Crab Creek (Washington) and Waha Lake (Idaho). The light broken line indicates the geographic separation between *lewisi* and *bouvieri* in the upper Columbia and upper Missouri river basins. Transfer of a Yellowstone cutthroat ancestor into the Bonneville Basin and into the Green River drainage of the Colorado River basin, and subsequent transfers into the South Platte (Arkansas) and Rio Grande river basins, probably occurred during the last glacial epoch. Dotted lines outline basins of internal drainage.

the San Luis Valley, Colorado (Rio Grande River basin), that existed more than 700,000 years ago. I can only assume that this ancient Rio Grande trout became extinct, and that the present Rio Grande cutthroat trout is not a direct descendent of the extinct form.

The coastal subspecies presently extends from the Eel River, California, to Gore Point, Kenai Peninsula, Alaska. This subspecies rarely occurs more than about 150 km inland from the Pacific coast.

The original distribution of the westslope subspecies also covered a vast area (Figure 2). It is native to the upper Kootenay River drainage of British Columbia and Montana (and tributaries in Idaho that are isolated by barrier falls from invasion by rainbow trout); the Pend Oreille–Clark Fork and Spokane–St. Joe river drainages of Montana and Idaho, above barrier falls located near the Washington–Idaho border; the Salmon and Clearwater rivers of the Snake River drainage, Idaho; the upper Columbia and Fraser river basins of British Columbia (a few isolated populations of mountain cutthroat trout, named *S. c. alpestris* by Dymond); the Lake Chelan drainage, Washington; and the John Day River drainage, Oregon. On the east side of the Continental Divide, *S. c. lewisi* is native to the South Saskatchewan River drainage of Montana and Alberta, and to the upper Missouri River drainage, Montana, where the original downstream distribution extended as far as Fort Benton. The native distribution in the upper Missouri drainage also includes the headwaters of the Milk and Marias rivers (Roscoe 1974).

The type locality of Yellowstone cutthroat trout is Waha Lake, Idaho, which was evidently isolated from the Snake River prior to the river's invasion by rainbow trout. The only other known site where a large-spotted Yellowstone-like cutthroat trout was found in the Columbia River basin below Shoshone Falls was in Crab Creek, near Ritzville, Washington, which was also isolated from the Columbia River. The now extinct Crab Creek cutthroat trout was named *Salmo eremogenes* by Evermann and Nichols (1909). I consider *eremogenes* to be a synonym of *S. c. bouvieri*. Evidently, the Yellowstone form of cutthroat trout was completely replaced by rainbow trout downstream from Shoshone Falls in the Columbia basin except for two sites isolated from contact with rainbow trout, Waha Lake and Crab Creek.

The recognition of a subspecies of cutthroat trout ("*Salmo mykiss gibbsi*") in the mid-Columbia basin by Jordan and Evermann (1896) was based on specimens of rainbow trout (Behnke 1981). The westslope cutthroat trout also was almost completely replaced by rainbow trout during the late glacial–postglacial period in the mid-Columbia basin below the barrier falls on the Kootenay, Pend Oreille, and Spokane rivers. Coexistence between indigenous westslope cutthroat and indigenous rainbow trout has persisted in the John Day River drainage, Oregon, and in the Salmon and Clearwater drainages of Idaho.

The greenback cutthroat trout in the upper South Platte and Arkansas river drainages and the Rio Grande cutthroat trout in headwaters of the Canadian River, New Mexico, have the easternmost natural distribution of the species. The Rio Grande cutthroat trout also has the southernmost documented occurrence, in southcentral New Mexico in tributaries to the Pecos River draining the slopes of the Sierra Blanca. Nineteenth-century reports of cutthroat trout in Texas and Mexico are unverified. Jordan and Evermann's (1896) listing of *S. c. pleuriticus* in the Little Colorado and White rivers of Arizona was based on Apache trout specimens (personal examination of the museum specimens). The Colorado River cutthroat trout was native to Arizona (now assumed extinct there) in the extreme northeast part of the state, where it occurred in streams of the Chuska Mountains of the San Juan River drainage.

Classification

All subspecies of cutthroat trout are not equal in terms of their length of evolutionary separation and degree of genetic divergence. The four major subspecies (*clarki, lewisi, bouvieri,* and *henshawi*) represent relatively large magnitudes of evolutionary divergence. The undescribed subspecies derived from *henshawi* in the Alvord and Whitehorse basins and in the Humboldt drainage of the Lahontan Basin, and the Paiute cutthroat trout *S. c. seleniris*, can be considered as minor subspecies derived from *henshawi*. The subspecies *utah, pleuriticus, stomias, virginalis, macdonaldi,* and the undescribed fine-spotted Snake River cutthroat trout can be considered as minor subspecies derived from the Yellowstone cutthroat trout.

Although there can be justifiable controversy concerning formal taxonomic recognition of the "minor" subspecies, the subspecies category in taxonomy historically has been more of a practical, "nonphylogenetic" category, simply used to designate "geographic races" of a species (Mayr 1969). A subspecies can be based on one original population (such as the Paiute trout) or on groups of populations in a defined geographical area whose members exhibit some detectable difference from all other subspecies (for example, the absence of spots on Paiute trout).

Rapid morphological change may result in diagnostic characters that can differentiate closely related subspecies such as *seleniris* from *henshawi* and *stomias, virginalis, utah,* and *pleuriticus* from each other and from *bouvieri*, even though no consistent genetic differences can be detected electrophoretically. The evolutionary time since the isolation of these subspecies (*seleniris* from *henshawi; stomias, virginalis, utah,* and *pleuriticus* from *bouvieri*) has been too brief for DNA changes coding for amino acid substitutions to occur in protein molecules that presently are amenable to electrophoretic analysis (which represent a very small part of fish genomes).

The practical aspect of classifying the species *Salmo clarki* into numerous geographic races recognized as subspecies concerns the conservation of biodiversity. The recognition and labeling of geographic races as unique forms of a species greatly facilitates state and federal management programs designed to protect and preserve natural diversity. Without its own unique subspecific designation, it is unlikely that the Paiute trout would have been listed as an "endangered species" and saved from extinction (Behnke 1981).

References

Behnke, R. J. 1968. A new subgenus and species of trout, *Salmo (Platysalmo) platycephalus*, from southcentral Turkey, with comments on the classification of the subfamily Salmoninae. Mitteilungen aus dem Hamburgischen Zoologischen Museum und Institut 66:1–15.

Behnke, R. J. 1981. Systematic and zoogeographic interpretation of Great Basin trouts. Pages 95–124 *in* R. J. Naiman and D. L. Soltz, editors. Fishes in North American deserts. Wiley, New York.

Cavender, T. M. 1986. Review of the fossil history of North American freshwater fishes. Pages 699–724 *in* C. H. Hocutt and E. O. Wiley, editors. The zoogeography of North American freshwater fishes. Wiley, New York.

Cavender, T. M., and R. R. Miller. 1982. *Salmo australis*, a new species of fossil salmonid from southwestern Mexico. Contributions from the Museum of Paleontology University of Michigan 26:1–17.

Evermann, B. W., and J. T. Nichols. 1909. Notes on the fishes of Crab Creek, Washington, with description of a new species of trout. Proceedings of the Biological Society of Washington 22:91–94.

Ferguson, A., and C. C. Fleming. 1983. Evolutionary and taxonomic significance of protein variation in the brown trout (*Salmo trutta* L.) and other salmonid fishes. Pages 85–99 *in* G. S. Oxford and D. Rollinson, editors. Protein polymorphism: adaptive and taxonomic significance. Academic Press, New York.

Gyllensten, U., and A. C. Wilson. 1987. Mitochondrial DNA of salmonids: inter- and intraspecific variability detected with restriction enzymes. Pages 301–317 *in* N. Ryman and F. Utter, editors. Population genetics and fishery management. University of Washington Press, Seattle.

Jordan, D. S., and B. W. Evermann. 1896. The fishes of North and Middle America. U.S. National Museum Bulletin 47 (part 1).

Kendall, A. R., and R. J. Behnke. 1984. Salmonidae: development and relationships. American Society of Ichthyologists and Herpetologists Special Publication 1:142–149.

Leary, R. F., F. W. Allendorf, S. R. Phelps, and K. L. Knudson. 1987. Genetic divergence and identification of seven cutthroat trout subspecies and rainbow trout. Transactions of the American Fisheries Society 116:580–587.

Loudenslager, E. J., and G. A. E. Gall. 1980. Geographic pattern of protein variation and subspeciation in cutthroat trout. Systematic Zoology 29:27–42.

Loudenslager, E. J., J. N. Rinne, G. A. E. Gall, and A. E. David. 1986. Biochemical genetics studies of native Arizona and New Mexico trout. Southwestern Naturalist 31:221–234.

Malde, H. E. 1965. The Snake River plain. Pages 255–264 *in* H. E. Wright and D. G. Frey, editors. The quaternary of the United States. Princeton University Press, Princeton, New Jersey.

Mayr, E. 1969. Principles of systematic zoology. McGraw-Hill, New York.

Murphy, T. C. 1974. A study of Snake River cutthroat trout. Master's thesis. Colorado State University, Fort Collins.

Rogers, K. L., and seven coauthors. 1985. Middle Pleistocene (Late Irvingtonian: Nebraskan) climate changes in southcentral Colorado. National Geographic Research 1:535–563.

Roscoe, J. W. 1974. Systematics of westslope cutthroat trout. Master's thesis. Colorado State University, Fort Collins.

Smith, G. R. 1981. Late Cenozoic freshwater fishes of North America. Annual Review of Ecology and Systematics 12:163–193.

Smith, G. R., and R. R. Miller. 1985. Taxonomy of fishes from Miocene Clarkia Lake beds, Idaho. Pages 75–83 *in* C. J. Smiley, editor. Late Cenozoic history of the Pacific Northwest. American Association for the Advancement of Science, Pacific Division, San Francisco.

Smith, R. H. 1984. Native trout of North America. Frank Amato Publications, Portland, Oregon.

Taylor, D. W., and G. R. Smith. 1981. Pliocene molluscs and fishes from northeastern California and northwestern Nevada. Contributions from the Museum of Paleontology University of Michigan 25: 339–413.

Trotter, P. C. 1987. Cutthroat, native trout of the West. Colorado Associated University Press, Boulder.

Vladykov, V. D. 1963. A review of salmonid genera and their broad geographical distribution. Transactions of the Royal Society of Canada (Series 3) 1:459–503.

Wilson, A. C., and nine coauthors. 1985. Mitochondiral DNA and two perspectives on evolutionary genetics. Biological Journal of the Linnean Society 26: 375–400.

American Fisheries Society Symposium 4:8–12, 1988

History of the Discovery of the Cutthroat Trout

Patrick C. Trotter and Peter A. Bisson

Weyerhaeuser Company, Technology Center, Tacoma, Washington 98477, USA

Abstract.—The cutthroat trout *Salmo clarki* was possibly the first trout to be recorded in writing from the New World, although the first scientific specimens were not collected until almost 300 years after its presence in western North America was noted. The initial record of a native trout was made in 1541 by a member of Coronado's army near the headwaters of the Pecos River in the Sangre de Cristo Mountains of New Mexico. In 1776, further mention of native western trout from Utah Lake and the Provo River was made by two Franciscan priests seeking to establish an overland route to the California missions from Santa Fe. Meriwether Lewis wrote of trout caught near the Great Falls of the Missouri River, as did other early Louisiana Territory explorers in different parts of the Great Basin. The first scientific specimens were taken by a young Hudson Bay Company physician, Meredith Gairdner, near Fort Vancouver, Washington, in late 1833 or early 1834. Two specimens were sent to Sir John Richardson in England, who named them in honor of Captain William Clark, coleader of the Lewis and Clark expedition. Owing to uncertainty over the identity of the Indian name for the river in which cutthroat trout were first collected, the Cathlapootl, the type locality of the species has remained conjectural for some time. Historical records, however, indicate that the Cathlapootl is the Lewis River in southwestern Washington.

The cutthroat trout *Salmo clarki* is the most widespread and polytypic salmonid native to western North America. Not surprisingly, it was encountered by a number of explorers of the region. Nearly all of the early explorers had little or no knowledge of fish systematics or zoogeography, but many of them kept handwritten records of their travels that later served as the basis of their expedition reports. These diaries, notes, and letters often contained references to unusual flora or fauna that were observed during the explorations. This paper recounts some of the very first references to cutthroat trout in the journals of the early explorers and concludes with a discussion of the circumstances surrounding the initial scientific description of the species. Additional references to early ichthyological observations and collections that include cutthroat trout can be found in Behnke (1979, 1981, and 1988, this volume) and Minckley et al. (1986).

Early Records

Coronado Expedition

An early record of native western trout is found in the chronicles of Pedro de Castaneda de Najera, a member of Coronado's army that explored what is now the southwestern USA from 1540 to 1542 searching for the legendary Seven Cities of Cibola. In the late spring of 1541, the army camped in a valley near the headwaters of the Pecos River in the Sangre de Cristo Mountains of northern New Mexico. The valley contained a large pueblo settlement called Cicuye, about which Castaneda wrote:

> Cicuye is located in a small valley between snowy mountain ranges and mountains covered with big pines. There is a little stream which abounds in excellent trout and otters. . . . [From Hammond 1940.]

The stream mentioned in the chronicle is most likely Glorieta Creek (J. V. Bezy, Pecos National Monument, personal communication), which flowed near the western edge of the pueblo complex but is now a sandy wash. The fish were almost certainly cutthroat trout because they occurred near the southernmost limit of the species' inland range, and no other trout are known to have lived in that region. (Smith and Miller 1986; Behnke 1988). The observation of trout in the headwaters of the Pecos River was apparently the first time any native trout had been recorded in writing from the New World.

Shortcut to the Missions

As western North America opened to European exploration, new routes were sought from the Great Plains to the Pacific Coast. One such route was a direct overland passage from Santa Fe (in present-day New Mexico) to the newly established Spanish missions in southern California. In 1776, two Franciscan priests, Father Silvestre Velezda Escalante and Father Francisco Anastasio Dominguez, left Santa Fe to locate a suitable trail. Their expedition traveled north of the Grand

Canyon into western Colorado, where they crossed the Colorado River and continued north to the White River. After turning westward, they eventually reached the Green River (which they called San Buenaventura) and continued west following the southern foothills of the Uinta Mountains to the Wasatch Range (northern Utah). A high pass led them to a river that Indians called the Timpanogus, which they followed to a large lake. At this point they abandoned their goal of finding a direct route to the missions, but before returning to Santa Fe they spent some time with the Indians that lived by the lake. The lake-dwellers relied heavily on aquatic resources, of which Father Escalante wrote:

> The Lake of the Timpanogitizes has great quantities of various kinds of food fish, geese, beaver and other amphibious animals which we had no opportunity to see. Round about it are a great number of these Indians who live on the abundant supply of fish in the lake. For this reason the Yutas Sabnaganas call them "fish eaters" [From Tanner 1936.]

The lake described by Father Escalante was Utah Lake, and the Timpanogus was the Provo River. Subsequent settlers of the area would also make great use of the fish populations in this drainage system and would seine cutthroat trout, some up to "a yard in length" (Suckley 1874) out of the lake by the ton.

Lewis and Clark Expedition

When Thomas Jefferson dispatched a small party of explorers in 1805 to examine the lands within the Louisiana Purchase, the expedition followed the Missouri River upstream to an area in present-day Montana called the Great Falls. The party halted to set up camp when they reached a fork in the river. With their Indian guide Sacajawea ill, Meriwether Lewis led a small contingent along the southernmost stream, which the two captains agreed was most likely the mainstem Missouri River. It was Lewis' contingent that first came upon the series of huge cataracts. A young member of the party named Silas Goodrich was an avid fisherman and often supplied the explorers with fish for dinner. Already he had caught two new species (probably mountain whitefish *Prosopium williamsoni* and Arctic grayling *Thymallus arcticus*) that had been recorded in the expedition journal. On the afternoon of the discovery of the Great Falls, Goodrich returned to camp with several trout in addition to whitefish. The trout were sufficiently different to cause Lewis to describe them in some detail in his daily record. Lewis' 1805 description was remarkably accurate and included the first references to a distinguishing cutthroat trout feature, the "cut" marks under the lower jaw.

> These trout are from 16 to 23 inches in length, precisely resemble our mountain or speckled trout in form and the position of their fins, but the specks on these are of a deep black instead of the red or goald of those common in the U' States. These are furnished with long teeth on the pallet and tongue and have generally a small dash of red on each side behind the front ventral fins; the flesh is of a pale yellowish red, or when in good order, of a rose red. [From Thwaits 1904.]

Expedition coleader William Clark went into even more detail in describing native fauna. In his notes there is a reference to cutthroat trout; however, at this early date it was easy to confuse the various kinds of salmon and trout, and it is not clear from this narrative whether the fishes he described were two color variations of the same species or two different species.

> Of the salmon–trout we observe two species differing only in colour. They are seldom more than two feet in length, and much narrower in proportion than the salmon or red char. The jaws are nearly of the same length and are furnished with a single series of small, subulate, straight teeth, not so long nor so large as those of the salmon. The mouth is wide, and the tongue is also furnished with small, subulate teeth, in a single series on each side; the fins are placed much like those of the salmon. One of the kinds, of a silvery-white colour on the belly and sides, and a bluish light brown on the back and head, is found below the Great Falls, and associates with the red char in little rivulets and creeks. It is about two feet eight inches long, and weighs ten pounds. The eye is moderately large, the pupil black, with a small admixture of yellow, and the iris of a silvery-white, and a little turbid near its border. . . . The fins are small in proportion to the size of the fish. Fins—*D*. 10–0; *P*. 13; *V*. 10; *A*. 12 — The other kind is of a dark colour on the back, and its sides and belly are yellow, with transverse stripes of dark brown; sometimes a little red is intermixed with these colours on the belly and sides toward the head. The eye, flesh, and roe are like those of the salmon. Neither this fish nor the salmon are caught with the hook, and we know not on what they feed. The white kind, found below the falls, is in excellent order when the salmon are out of season and unfit for use. [From Thwaits 1904.]

Part of this description was taken from Clark's handwritten journal, which included a reference to "white salmon trout" of large size taken near the Great Falls, not of the Missouri River but of the Columbia River (possibly Celilo Falls). From the sketch drawn in the journal and from additional notes it would appear that "white salmon trout" were steelhead *Salmo gairdneri*, while the smaller fish not illustrated in Clark's journal but elsewhere referred to as "dark salmon trout" was later taken by Richardson (1836) to be the cutthroat trout.

Other Early Explorers

A number of other 19th century explorers mentioned trout in diaries of their travels around western North America. From 1808 to 1812, David Thompson, a trader for the British Northwest Fur Company, made notes of several catches of trout from the northern Rocky Mountains. Included in Thompson's notes were records of trout from Mission Creek in Montana; a tributary of Kootenay River near Rexford, Montana; the Kootenay River between Bonner's Ferry and Lake Pend Oreille, Idaho; and Windemere Lake, British Columbia (White 1950). Early explorers of the Oregon Trail system from 1812 to 1821 relied heavily on trout and other fishes for subsistence during the winter in Montana and Idaho, according to the chronicles of Robert Stuart (Rollins 1935). Notes from the Great Basin explorations of Peter Skene Ogden included Indian accounts of large "salmon" from the Truckee River (Cline 1963), a reference to Lahontan cutthroat trout *Salmo clarki henshawi* that ascended this tributary of Pyramid Lake to spawn. Further accounts of large Pyramid Lake cutthroat trout are found in the writings of John C. Fremont (1845), who also gave excellent descriptions of the dams used by the Indians to trap adult fish on their way upstream.

Almost without exception, these early records of cutthroat trout refer to the populations as being extremely abundant. For example, in the journals of John Kirk Townsend, who traveled in a party under the leadership of Nathaniel Wyeth to deliver trade goods to mountain fur trappers, there appears a passage describing the abundance of trout in the Bear River of Wyoming.

> In this little stream, the trout are more abundant than we have yet seen them. One of our *sober* men took, this afternoon, upward of thirty pounds. These fish would probably average fifteen or sixteen inches in length, and weight three-quarters of a pound; occasionally, however, a much larger one is seen. [Townsend 1839.]

Edward Hewitt, while a boy of 15, wrote of accompanying his father in 1881 to the recently established Yellowstone National Park where he often supplied trout from the Yellowstone River for the camp. His writings included a story of fishing a small stream one afternoon to feed 40 soldiers.

> The stream was just alive with trout, which seemed to run from three to four pounds apiece. I was not sorry to quit as I was really tired out. For once I had caught all the fish I could take out in one day. There must have been between four hundred and fifty and five hundred pounds of cleaned trout, but the soldiers polished them off in two meals. . . . [Hewitt 1948.]

First Scientific Description

Although the coastal cutthroat trout *Salmo clarki clarki* may have been one of the last forms of the species to be encountered by early explorers of the region, it was the first form to be given official scientific recognition. In the first half of the 19th century, naturalists roamed the New World seeking unusual flora and fauna for scientific study. Many of the specimens were shipped back to Europe where they were described in books or journals such as the *Publications of the Royal Society of London*. William Jackson Hooker, director of the Kew Botanic Garden, encouraged a number of talented young naturalists, including Charles Darwin, to travel abroad and bring back new discoveries. At almost the same time that Darwin departed on his fateful scientific voyage, Hooker received word that physicians were needed to help fight an outbreak of "intermittent fever" (malaria) at an outpost of the Hudson Bay Company near the Columbia River in western North America. Acting on the advice of Sir John Richardson, Hooker recommended W. F. Tolmie and Meredith Gairdner. The two young doctors sailed from London in September 1832, and arrived at Fort Vancouver on the lower Columbia River in late April 1833. Although Gairdner was highly respected by Richardson, Hooker, and local naturalist David Douglas, he had very little time to travel the nearby countryside searching for interesting plants and animals. After only 11 months at Fort Vancouver he contracted symptoms of a serious respiratory infection, possibly tuberculosis. In late March 1834, Gairdner left Fort Vancouver to convalesce, but the illness was too severe. On March 26, 1837, he died in Hawaii at age 28 (Harvey 1945).

One of Gairdner's projects at Fort Vancouver had been to study the salmon of the Columbia River. In 1834, a container with six specimens of salmon and two small trout was sent to Richardson to be described in his comprehensive treatise *Fauna Boreali-Americana* (Richardson 1836). The specimens were wrapped, preserved in alcohol, and soldered in a tin case that was placed in a protective cask. Despite these precautions, the container was damaged during the long sea voyage to England, and much of the preservative was lost. According to Richardson ". . . six specimens of salmon were incorporated into one mass by the continued motion of the vessel. The other fish, being of a smaller size, less oily, and perhaps more indurated by longer immersion in spirits, arrived in better condition." Thus, in 1836, Richardson named the new trout *Salmo clarkii* ". . . as a tribute to the memory of Captain Clarke, who notices it in the narrative prepared by him of the proceedings of the Expedition to the Pacific, of which he and Captain Lewis had a joint command, as a dark variety of Salmon-trout. . .."

Richardson's description of the coloration of the specimens reflects the poor condition in which they had arrived.

> Back generally brownish purple-red, passing on the sides into ash-grey, and into reddish-white on the belly. Large patches of dark purplish-red on the back. Dorsals and base of the caudal ash-grey, end of caudal pansy-purple. Back, dorsal, and caudal studded with small semilunar spots. A large patch of arterial-red on the opercle and margin of the preopercle. Pectorals, ventrals, and anal greyish-white tinged with rose-red. [Richardson 1836.]

The type locality of the cutthroat trout has remained somewhat of a mystery. In his original description, Richardson stated, "Dr. Gairdner does not mention the Indian name of this trout, which was caught in the Katpootl, a small tributary of the Columbia on its right bank. . . ." At the time of the Lewis and Clark expedition, all lower Columbia River tributaries had small fishing villages that marked the better fishing spots, but the malaria epidemics of 1829–1832 decimated as many as 90% of the Wappato area Indians living in the vicinity of Fort Vancouver (Jones 1972). The available evidence, however, points to the Lewis River as being the stream in which the first scientific specimens were collected. One of the first records of the stream is found in the journals of Gabriel Franchere, who arrived at the mouth of the Columbia River with John Jacob Astor in 1811. While investigating a report of white settlers in the area upstream from the mouth of the Columbia, Franchere wrote:

> A.M. of May 6. Entered a "little river" to the village of Kalama, where young chief "Keassen" was residing at the time. They then proceeded to the large village of "Katlapoutle", situated at the mouth of a small river that seemed to flow down from a snow covered mountain. . .." [From Franchere 1967.]

The snow-covered mountain was Loo-Wit (Mount St. Helens), and the river was most likely the Lewis. An early map of the region drawn by Alexander Ross between 1810–1814 has the name "Cattla puttle tribe" inscribed near what is apparently the Lewis River drainage. Hodge (1907) reported the following information regarding an Indian settlement at the mouth of the river:

> *Cathlapotle* ("people of Lewis R."). A Chinookian tribe formerly living on the lower Lewis River—on the SE side of the Columbia in Clarke Co., Washington. . . . Other names and spellings listed: Cath-lah-poh-tle, Cathlapootle, Cathlapoutles. . .. The people of this village and perhaps all others of this particular group residing along the Lewis R. were wiped out completely by the "intermittent fever" outbreak. . . .

The first European settlement along the river was made in 1845 by Adolphus Lee Lewis, an Englishman employed by the Hudson Bay Company (Parsons 1983). When George McClellan conducted his Pacific Railroad survey in 1853, he went overland to the river from the U.S. military post at Fort Vancouver. McClellan then referred to the river as both the Cathlapootle and the Lewis in his diary (McClellan 1853). Two weeks later the Dryer party followed McClellan's trail enroute to the first ascent of Mount St. Helens. It was in Dryer's account of this journey, published in his newspaper, the *Portland Oregonian*, that he referred to the same river as the Lewis. After that, the name Lewis River was generally used.

The actual location on the Lewis River where Gairdner's specimens were taken is unknown. They could have been captured near the confluence of the Lewis and Columbia rivers if the Lewis River had been reached by canoe from Fort Vancouver, or they could have been collected from an upstream reach if an overland route through Chelatchie Prairie had been tried. It is also not known whether Gairdner himself collected the fish or whether the specimens were brought to him. In any case, Gairdner's efforts to obtain specimens in the face of formidable difficulty were not scientifically unrewarded; a variety of downy woodpecker *Dendrocopos pubescens gairdneri* was named after him, as well as the rainbow trout.

Acknowledgments

Many people assisted in locating archival references to trout in the American west. In particular, we are grateful to the Oregon Historical Society, the Oregon State University library, Pecos National Monument (U.S. National Park Service), the Clark County (Washington) Museum, Cascade County (Montana) Historical Society, the Thomas Gilcrease Museum, Pacific Power and Light Company, the American Philosophical Society Library, Washington State Historical Society, Utah State Division of State History, and the Fort Vancouver National Historic Site (U.S. National Park Service). Robert J. Behnke kindly reviewed the manuscript. We would like to dedicate this paper to the late Professor Roland E. Dimick of Oregon State University, whose interest in the knowledge of the natural history of cutthroat trout was a source of inspiration.

References

Behnke, R. J. 1979. The native trouts of the genus *Salmo* of western North America. Report to U.S. Fish and Wildlife Service, Denver, Colorado.

Behnke, R. J. 1981. Systematic and zoogeographic interpretation of Great Basin trouts. Pages 95–124 *in* R. J. Naiman and D. L. Soltz, editors. Fishes in North American deserts. Wiley, New York.

Behnke, R. J. 1988. Phylogeny and classification of cutthroat trout. American Fisheries Society Symposium 4:1–7.

Cline, G. G. 1963. Exploring the Great Basin. University of Oklahoma Press, Norman.

Franchere, H. C., editor and translator. 1967. Adventures at Astoria 1810–1814 by Gabriel Franchere. University of Oklahoma Press, Norman.

Fremont, J. C. 1845. Report of the exploring expedition to the Rocky Mountains in the year 1842 and to Oregon and North California in the year 1843–1844. U.S. Senate, 28th Congress, 2nd Session, Executive Document 174. (Printed by Gale and Seaton, Washington, D.C.)

Hammond, G. P. 1940. Narratives of the Coronado expedition, volume 2. University of New Mexico Press, Albuquerque.

Harvey, A. G. 1945. Meredith Gairdner: doctor of medicine. British Columbia Historical Quarterly 9:89–111.

Hewitt, E. R. 1948. A trout and salmon fisherman for seventy-five years. Charles Scribner's Sons, New York. (Reprint 1966. Abercrombie and Fitch, New York.)

Hodge, F. W. 1907. Handbook of the American Indians. Smithsonian Institution, Bureau of American Ethnology, Bulletin 30, Washington, D.C.

Jones, R. F. 1972. Wappato Indians of the lower Columbia River valley: their history and prehistory. Private printing, Vancouver, Washington.

McClellan, G. B., Sr. 1853. Diaries, May 20 to December 11, 1853, and July 15 to September 30, 1853. Library of Congress, McClelland Papers, Manuscript Division, Washington, D.C.

Minckley, W. L., D. A. Hendrickson, and C. E. Bond. 1986. Geography of western North American freshwater fishes: descriptions and relationships to intracontinental tectonism. Pages 519–613 *in* C. H. Hocutt and E. O. Wiley, editors. The zoogeography of North American freshwater fishes. John Wiley and Sons, New York.

Parsons, M., editor. 1983. Clarke County, Washington Territory. Washington Publishing, Portland, Oregon.

Richardson, J. 1836. Fauna Boreali-Americana. Richard Bently, London. (Reprint 1978. Arno Press, New York.)

Rollins, P. H., editor. 1935. The discovery of the Oregon Trails: Robert Stuart's narratives. Scribner's, New York.

Smith, M. L., and R. R. Miller. 1986. The evolution of the Rio Grande basin as inferred from its fish fauna. Pages 457–485 *in* C. H. Hocutt and E. O. Wiley, editors. The zoogeography of North American freshwater fishes. Wiley, New York.

Suckley, G. 1874. On the North American species of salmon and trout. Pages 91–160 *in* Report of the Commissioner for 1872 and 1873. U.S. Commission of Fish and Fisheries, Washington, D.C.

Tanner, V. M. 1936. A study of the fishes of Utah. Proceedings of the Utah Academy of Sciences, Arts and Letters 13:155–183.

Thwaits, R. G., editor. 1904. Original journals of the Lewis and Clark expedition. Dodd and Mead, New York.

Townsend, J. K. 1839. Narrative of a journey across the Rocky Mountains to the Columbia River. (Reprinted 1978. University of Nebraska Press, Lincoln.)

White, M. C., editor. 1950. David Thompson's journals relating to Montana and adjacent regions 1808–1812. Montana State University Press, Missoula.

American Fisheries Society Symposium 4:13–24, 1988

Ecology, Status, and Management of the Yellowstone Cutthroat Trout

JOHN D. VARLEY

U.S. National Park Service, Post Office Box 168
Yellowstone National Park, Wyoming 82190, USA

ROBERT E. GRESSWELL

U.S. Fish and Wildlife Service, Post Office Box 184
Yellowstone National Park, Wyoming 82190, USA

Abstract.—The Yellowstone cutthroat trout *Salmo clarki bouvieri* is the most abundant and widely dispersed subspecies of inland cutthroat trout. The historic range of the subspecies included the Yellowstone River drainage in Montana and Wyoming and portions of the Snake River drainage in Wyoming, Idaho, Nevada, Utah, and perhaps Washington. Introductions of nonnative fishes, environmental degradation, and human exploitation have severely reduced this range. The subspecies is estimated to presently exist in a genetically pure form in about 85% of the historic lake habitat but only in about 10% of the estimated original stream range. Additionally, Yellowstone cutthroat trout populations have been established outside the historic range in at least seven western states and two Canadian provinces. High landing rates and angler success, coupled with the fish's relatively large mean length, make the Yellowstone cutthroat trout a valuable recreational resource; however, vulnerability to angling has often led to overharvest of wild stocks. Current management of the Yellowstone subspecies includes wild trout programs for native and introduced populations and maintenance stocking from hatchery broodstocks. Genetic integrity, habitat management, and special angling regulations are important elements that influence the success of these programs. The status of the Yellowstone cutthroat trout is encouraging, but future management must emphasize protection of the full range of genetic diversity.

With the exception of the coastal cutthroat trout *Salmo clarki clarki,* there is no other cutthroat trout subspecies as abundant as, nor one that covers a greater geographic range than the Yellowstone cutthroat trout *Salmo clarki bouvieri*. To most people, however, the Yellowstone subspecies is synonymous with the widely transplanted and propagated race from Yellowstone Lake, which, for over 50 years, was the most commonly introduced cutthroat trout in the world. In terms of subspecies survival or potential sport-fish value, it is ironic that the race used for massive distribution was the form of Yellowstone cutthroat trout least adapted for introduction into different environments and competitive fish communities.

Since the subspecies was isolated during deglaciation (6,000–8,000 years ago), Yellowstone cutthroat trout have evolved from an apparently homogeneous population into several distinctive races able to take advantage of varying environmental conditions (Behnke 1979). The Yellowstone Lake cutthroat trout, which is a highly adapted example of this diversity, is but one of a large assemblage of *S. c. bouvieri* populations with distinctive life history characteristics.

Reproductive isolation between populations of Yellowstone cutthroat trout has magnified behavioral differences and given rise to wide divergence in coloration, morphology (Cope 1957a; Bulkley 1963; Behnke 1979), and biochemical-genetic factors (Loudenslager and Gall 1981). Several authors have suggested significant racial differences between populations occurring within the same water. Cope (1957a) presented evidence that several tributaries to Yellowstone Lake maintained discrete races. Bulkley (1963) noted significant differences in spotting, hyoid teeth counts, and coloration between isolated populations of cutthroat trout in the Yellowstone Lake drainage. Liebelt (1968) also found discrete populations by using analysis of serum antigens.

Racial differences among Yellowstone cutthroat trout within a given water body probably resulted from reproductive isolation caused by homing instincts that return the fish to natal streams. Strong homing tendencies to specific spawning streams, both within a year and in subsequent years, have been described for fish in Yellowstone Lake (Ball 1955; Cope 1957b; Mc-

Cleave 1967; Jahn 1969) and in Strawberry Reservoir, Utah (Platts 1959). LaBar (1971) reported that, after displacement, individual fish returned to a specific redd. Working with nonspawning fluvial cutthroat trout, Miller (1954) reported the fish homed to previously occupied territories within a stream. Cope (1957b) believed that Yellowstone Lake cutthroat trout homed to the streams in which they were born and even to the precise natal area, but the latter has never been experimentally studied. On tributaries to Yellowstone Lake, research with marked cutthroat trout suggested that straying rates between streams were low, generally making up less than 2% of the spawning population (Ball 1955; Jones et al. 1982).

In this paper, we attempt to define some of the known diversity within the subspecies, demonstrate the value of preserving the remaining diversity, and suggest how this diversity might be important for sport fisheries in the future.

Range

The historic range of the subspecies included the Yellowstone River drainage in Montana and Wyoming and portions of the Snake River drainage in Wyoming, Idaho, Nevada, Utah, and, perhaps, Washington (Behnke 1979 and 1988, this volume; Figure 1). The Yellowstone subspecies originally circumscribed the range of a purported unnamed subspecies, the fine-spotted Snake River cutthroat trout, and overlapped the range of the coastal cutthroat trout. Introductions of nonnative fishes, environmental degradation, and human exploitation severely reduced this range when Europeans settled the western USA (Figure 1). Stocking activities by agencies and private individuals expanded the original range somewhat and established self-sustaining populations outside the original range. These transplants, however, never restored the Yellowstone cutthroat trout to historic proportions.

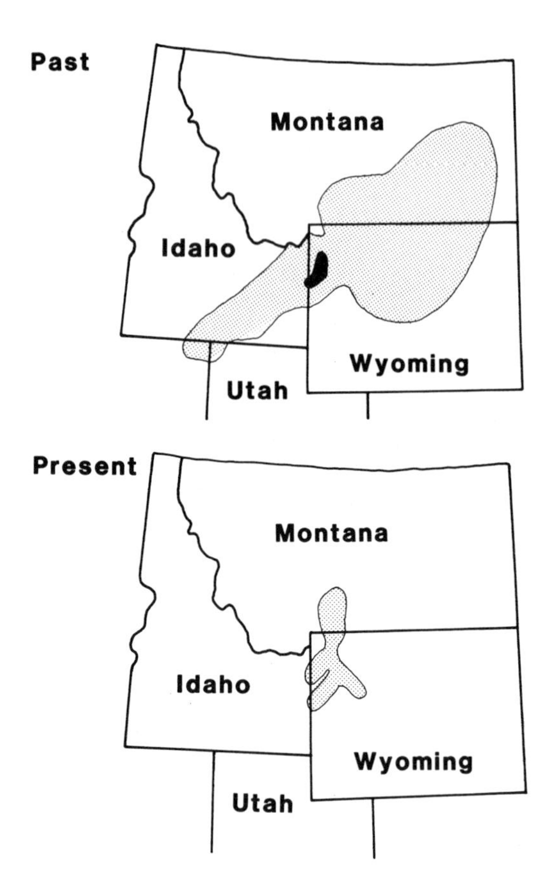

FIGURE 1.—Probable past (pre-1880s) and present distributions of native Yellowstone cutthroat trout *Salmo clarki bouvieri* (light shading). Dark shading in Wyoming is the probable distribution of the unnamed fine-spotted Snake River cutthroat trout.

Within the historic range, the subspecies is presently estimated to exist in pure form in about 38,500 hectares of lakes and 2,400 km of streams. This is an encouraging estimate for lake populations because it means about 85% of the minimum estimate of the original habitat (44,500 hectares) is still occupied. By contrast, only about 10% of the estimated original stream range of about 24,000 km remains inhabited.

It is difficult to define the current expanded range of the Yellowstone cutthroat trout. Over 818 million eggs and fry were shipped from Yellowstone National Park (Varley 1979), and there were several other egg-taking and -shipping facilities outside the park. Waters in over one-half of the 50 United States, most of the Canadian provinces, in several other countries received shipments of Yellowstone cutthroat trout between 1899 and 1957. Virtually all of these fish were from the highly specialized, lacustrine Yellowstone Lake stock. In contrast to the success of brown trout *Salmo trutta,* rainbow trout *S. gairdneri,* and brook trout *Salvelinus fontinalis* transplanted throughout North America, the present scarcity of Yellowstone cutthroat trout outside their native range suggests an inherent lack of adaptability.

Despite the widespread failure of these transplants, Yellowstone cutthroat trout populations have been established outside the original range in Montana, Wyoming, Idaho, Nevada, Washington, Utah, Oregon, Colorado, and probably some other western states (Sigler and Miller 1963; Carlander 1969; Baxter and Simon 1970; Brown 1971; Simpson and Wallace 1978). Introduced populations also exist in British Columbia, Alberta, and perhaps Quebec (Scott and Crossman 1973). Most of the successful establishments occurred when this subspecies was transplanted into small alpine or subalpine lakes that were fishless prior to the introductions. Confusion concerning genetic purity precludes an accurate estimate of the number of populations or their extents. Throughout their present range, Yellowstone cutthroat trout commonly hybridize with other subspecies of cutthroat trout, with rainbow trout, and, in some cases, with golden trout *Salmo aguabonita,* and hybridization has been blamed for the decline in the genetic integrity of the Yellowstone cutthroat trout and for the demise of other rarer trout species and subspecies (Behnke and Zarn 1976).

In addition to existing wild populations, there are several hatchery brood stocks of Yellowstone cutthroat trout. The states of Wyoming (South Paintrock Creek race) and Montana (McBride Lake race) culture brood stocks for stocking in (primarily) mountain lakes and reservoirs. Although a hatchery brood stock is no longer maintained in Idaho, eggs are collected annually from Yellowstone cutthroat trout in tributaries to Henrys Lake, Idaho.

Status

Several factors have contributed to the success of the Yellowstone subspecies within its native range. At present, remaining populations are located in remote, inaccessible headwater areas that did not have major contact with Europeans until relatively recently. The region was not extensively explored before the late 1860s and has never been densely inhabited by humans. The earliest transplants of nonnative fishes into the area did not occur until 1889 (Varley 1981). Delayed and incomplete human habitation of these remote mountain areas served to buffer Yellowstone cutthroat trout from human abuse.

It is also notable that most of the mountain wilderness harboring Yellowstone cutthroat trout was retained in public ownership as parks or forest reserves. Yellowstone National Park was established in 1872, just a few years following its initial exploration. The first national forest in the USA, initially known as the Yellowstone Park Forest Reservation and now as the Shoshone National Forest, was created in 1891. Several parks, refuges, and a number of national forests followed after the turn of the century. Given this kind of protection, lakes and streams generally did not experience the damaging human activities that commonly affected other, less protected American fisheries.

Despite some abuses and unfortunate transplants, the early stewards of the fish, wildlife, parks, and forests had some ideas that were imaginative and served to protect populations and gene pools. Early in this century, Yellowstone National Park administrators and the U.S. Commissioner of Fisheries elected not to intermix salmonid species, and each major drainage in the park was reserved for a different salmonid (Jordan 1891). Although there is evidence that this decision was based more on the desire to restrict the spread of infamous Yellowstone Lake fish parasites than to preserve the genetic integrity of Yellowstone cutthroat trout, it was a fortunate decision for the subspecies. In addition to preserving Yellowstone cutthroat trout genotypes, this policy also retarded subsequent introductions, such as stocking brook trout in waters with native Yellowstone cutthroat trout populations. In Yellowstone National Park, the introduction of brook trout has nearly always resulted in the disappearance of Yellowstone cutthroat trout.

Delayed advent of human exploitation further served to protect Yellowstone cutthroat trout populations. Given a proven vulnerability to angling, survival of this subspecies would have been doubtful if it had been subjected to commercial, subsistence, and sport fishing typical of fisheries across the North American continent during the 18th and 19th centuries. In the greater Yellowstone area, unfished populations existed until the late 1800s; the first notable signs of exploitation on Yellowstone Lake and the Yellowstone River, for example, were not evident until about 1919 (Albright 1920). For most populations, overfishing apparently did not begin until the 1950s and 1960s.

The status of the Yellowstone cutthroat trout inside and outside its native range is encouraging. Perhaps the greatest danger that the subspecies faces is that the number of ecotypes now existing (hence the full range of genetic variability) may continue to decline through both natural events and human activity.

Life History

Many aspects of the life history of Yellowstone cutthroat trout have been studied, and they are briefly summarized here. An attempt has been made to reference only that information relating to genetically pure populations existing within their native range; however, information about several established transplanted populations outside of this range is also incorporated.

Yellowstone cutthroat trout are known to spawn over a 7-month period (actual spawning time depends on latitude, altitude, water temperature, and runoff conditions). In the greater Yellowstone area, spawning can occur from the latter part of April through early August. Spawning migrations from lakes in Yellowstone National Park are often affected by elevation. For example, at Trout Lake (elevation, 2,121 m), peak migration takes place around May 20; in tributaries to Yellowstone Lake (elevation, 2,357 m), the peak is variable and occurs between late May and early July; and in Sylvan Lake (elevation, 2,565 m), the peak occurs during the latter part of July. Bureau of Fisheries spawning records (archive documents, Bozeman National Fish Hatchery, Montana), covering the early part of this century, show egg collections as late as mid-September for Clear Creek (one of the coldest tributaries to Yellowstone Lake), but this stock is apparently extinct.

Yellowstone cutthroat trout transplanted outside the original range have also shown wide variation in time of spawning. The senior author has observed cutthroat trout migrating upstream in a tributary to Flaming Gorge Reservoir, Utah–Wyoming (elevation, 1,829 m), as early as mid-March. Spawning migrations in Strawberry Reservoir, Utah (elevation, 2,300 m), occurred early in June (Platts 1959). Fleener (1952) observed gravid females as early as April 12 and as late as August 16 in various parts of the Logan River drainage, Utah (elevation range, 1,525–2,286 m).

Mature Yellowstone cutthroat trout, whether residents of lakes or streams, spawn in fluvial environments. Lacustrine spawning has never been documented; however, in the first 50 years of this century, spawning-trap workers operating weirs on tributaries to Yellowstone Lake believed that trout spawned in the lake shoals during years of high spring runoff (Fromm, circa 1941).

Yellowstone cutthroat trout are known to exhibit four different migratory-spawning patterns.

(1) Fluvial populations disperse locally for spawning in the area of their home range within a stream. Yellowstone cutthroat trout in the Yellowstone River from Yellowstone Lake outlet to the Upper Falls of the Yellowstone River display this type of migration. Resultant fry may move upstream or downstream or may hold position in the area of natal gravel.

(2) Fluvial–adfluvial populations migrate from larger rivers to smaller tributaries to spawn. The Yellowstone River below the Lower Falls and the Lamar River, which have limited suitable spawning habitat in the main stems, are examples. Progeny often spend 1, 2, or 3 years in the tributaries before migrating to the larger river to reside.

(3) Lacustrine–adfluvial populations of Yellowstone cutthroat trout reside in lakes and ascend tributaries to spawn. The majority of the lake populations display this behavior. Fry typically migrate to the lake shortly after emergence from the gravel. Where habitat and environmental conditions are adequate, some fry may spend one or more years in the stream prior to emigration. Welsh (1952) hypothesized that fry remaining in residency for 1 or 2 years were those that emerged in the headwaters (2-year residence) or middle reaches (1-year residence).

(4) Allacustrine populations migrate downstream through a lake outlet to spawn. This is the rarest type of migratory behavior exhibited by Yellowstone cutthroat trout. Progeny follow different behavioral cues and move upstream to the lake following emergence. This behavior is heritable (Raleigh and Chapman 1971; Bowler 1975). Yellowstone Lake and Heart Lake (Yellowstone National Park) have significant allacustrine stocks. In Heart Lake, the outlet group spawns about 1 month (mid-July) after adfluvial spawners (mid-June); thus, they are reproductively isolated from each other in both space and time.

In the outlet from Yellowstone Lake, fluvial cutthroat trout from the Yellowstone River spawn in the same area and at about the same time as allacustrine fish (Ball and Cope 1961). Although some fry apparently rear in the spawning area after emergence, fry migrations upstream to Yellowstone Lake and downstream in the Yellowstone River have been observed.

In spring, older and larger spawners migrate first (Ball and Cope 1961; Jones et al. 1984). Age, length, weight, and condition factor of individual spawners decline as the run progresses (Jones et al. 1984). As with other fish species (Briggs 1953), there is evidence that older and larger Yellowstone cutthroat trout migrate farther upstream than smaller individuals (Cope 1957b; Dean and Varley 1974).

Spawners in Yellowstone Lake tributaries spend between 6 and 25 d in the stream. Males generally migrate earlier than females, and they remain longer (Ball and Cope 1961). Exploited populations show a proportion of initial spawners as high as 99% with few repeat spawners (Ball and Cope 1961); however, recent studies at Clear Creek, a tributary to Yellowstone Lake, indicated that at least 26% of the spawners had spawned previously (Jones et al. 1982). During an era when angler harvest exceeded the estimated maximum sustained yield (Benson and Bulkley 1963), the average age of spawners ascending Clear Creek was 3.9 years. Mean age has increased to 5.0 years since harvest was reduced by special regulations (Gresswell and Varley 1988, this volume). Even in lightly exploited or unexploited populations, however, repeat spawning occurs in alternate years more frequently than in consecutive years (Jones et al. 1984). In the Rocky Mountains, alternate-year spawning is more common in populations found at higher altitudes (over 2,100 m) than in those at lower elevations; this phenomenon appears to be related to growth rates, degree of parasitic infection, or other factors (Ball and Cope 1961).

High postspawning mortality has been reported in the past. Ball and Cope (1961) estimated an average instream mortality of 48.1% in five Yellowstone Lake tributaries between 1948 and 1953. Estimates for instream and between-year postspawning mortality for Yellowstone cutthroat trout in Strawberry Reservoir, Utah, were somewhat higher (Platts 1959). Recent research on a Yellowstone Lake stock failed to substantiate the high instream mortality rate reported in the earlier studies. Instream spawning mortality in Clear Creek averaged 12.9% for five recent sample years (1977–1979, 1983, and 1984; Jones et al. 1985). Excessive harvest of adults and reduction of multiple-age spawning stocks to single-age runs have been shown to be determinants associated with recruitment instability in Yellowstone Lake (Gresswell and Varley 1988).

Standing crops of Yellowstone cutthroat trout in 23 mountain streams in Wyoming and in Yellowstone National Park ranged between 7 and 145 kg/hectare and averaged 48 kg/hectare (Binns and Eiserman 1979; Yellowstone National Park, unpublished data). Less is known about standing crop in lakes. Two small subalpine lakes in Yellowstone National Park had estimated standing crops of catchable-size (>200 mm total length, TL) Yellowstone cutthroat trout of 23 and 54 kg/hectare (Jones et al. 1979). Estimates of cutthroat trout biomass in Yellowstone Lake, which supports the largest inland cutthroat trout population in the world, have historically been difficult because of the large size of the lake (35,400 hectare) and the number of cutthroat trout. In 1979, several methods were used to estimate the catchable stock. Resulting estimates were 1–4 million Yellowstone cutthroat trout greater than 350 mm TL, but confidence limits were wide (Jones et al. 1980). Expansion of these figures would suggest a standing crop of 12–43 kg/hectare.

Habitat Requirements

Population estimates and habitat characteristics from 13 cutthroat trout streams in Wyoming indicated that the following nine environmental attributes were instrumental in explaining over 90% of the variation in standing crops (Binns and Eiserman 1979): late summer streamflows, annual streamflow variation, water velocity, trout cover, stream width, eroding stream banks, stream substrate, nitrate–nitrogen concentration, and maximum summer water temperature.

Although the Yellowstone cutthroat trout is often considered a "headwater" subspecies, its native habitat has been far more diverse than it is now. Historically, its elevational range extended from about 275 m (Waha Lake, Idaho) to at least 2,590 m (Lake Eleanor, Yellowstone National Park). The current range varies from about 1,300 m (Yellowstone River, Montana) to over 3,200 m (Wind River Mountains, Wyoming), although some introduced populations may occur at lower elevations. Populations were (and still are) found in waters ranging in size from small beaver impoundments to large lakes (e.g., Yellowstone Lake, 35,400 hectares). Historically, Yellowstone cutthroat trout were common in large western rivers such as the Snake River above Shoshone Falls, Idaho (mean annual flow, 156 m^3/s), and the Yellowstone River at Miles City, Montana (mean annual flow, 321 m^3/s); presently they survive only in headwaters of these rivers.

The reported preference of Yellowstone cutthroat trout for small headwater streams may be due to the number of populations that have been found in streams as small as 1 m wide, with flows as low as 0.06 m^3/s. In alpine and subalpine climates, these populations have survived at an environmental extreme, overwintering in a trickle of water for more than 8 months in extreme cold and severe ice conditions.

Yellowstone cutthroat trout are adapted to cold water; winter observations on Yellowstone Lake showed that Yellowstone cutthroat trout actively fed at 0–4°C beneath a 1-m thick ice cover (Jones et al. 1979). In higher subalpine streams, viable, and in some cases robust, populations exist in streams with summer maxima in the range of 5–8°C (Gregg Fork Bechler River, Sedge and Bear creeks, all in Yellowstone National Park; Jones et al. 1979).

Water temperatures between 4.5 and 15.5°C appear to be optimum for the subspecies (Carlander 1969); however, populations have been known to exist in waters heated geothermally to at least 27°C in Yellowstone National Park (Alum and Witch creeks, tributaries to Pelican Creek, and Yellowstone Lake tributaries at West Thumb Geyser Basin), and temperatures exceeding 26°C were probably common in various parts of the historical range (Yellowstone River near Miles City, Montana; Bighorn River drainage, Wyoming; and Snake River near Shoshone Falls, Idaho).

The subspecies is currently found in waters that display chemical characteristics ranging in pH from about 5.6 (Fern Lake, Yellowstone National Park) to over 10.0 (McBride Lake, Yellowstone National Park), with total dissolved solids between about 10 and 700 mg/L. In Yellowstone National Park, where geothermal waters often create unusual chemical conditions, Yellowstone cutthroat trout are not found in waters where the pH ranges from 4.0 to 5.0, even if other conditions, such as temperature, seem favorable. Much greater chemical variation undoubtedly occurred in the historic range. For example, certain surface waters in the Bighorn River drainage (Wyoming) commonly exceed 2,000 mg/L total dissolved solids (U.S. Geological Survey, unpublished data).

Yellowstone cutthroat trout migrate when threshold water temperatures approach 5°C (optimum 10°C) and streamflows subside from spring peaks (Ball and Cope 1961; Jones et al. 1980). Upstream migrations most often occur during daylight hours as temperatures rise and flows decrease. Postspawning downstream emigrations are nocturnal during the early portions of the season when flows are high, but are concentrated during daylight hours in the later part of the migration. Cope (1956) described nocturnal upstream spawner migration in Arnica Creek (tributary to Yellowstone Lake) as being normal; however, Carlander (1969) observed that nocturnal migration was unusual for salmonid species. Recent studies (1973–1983) on three other tributaries to Yellowstone Lake failed to show significant nocturnal movement (Jones et al. 1980, 1983, 1984).

Streams selected for spawning are commonly low-gradient (up to about 3%), perennial streams, with groundwater and snow-fed water sources. Use of intermittent streams for spawning is not well documented, but spawning has been noted recently in some intermittent tributaries to Yellowstone Lake (Jones et al. 1986). In these seasonal waters, numerous throughout the range of the subspecies, spawning typically takes place during spring runoff (May–June), and emergent fry migrate to a lake or larger stream in July–August before the natal stream is dry. Erman and Hawthorne (1976) described the importance of an intermittent stream to the maintenance of a rainbow trout population, especially as it related to competition with brook trout. We suspect that the ability of the Yellowstone cutthroat trout to use intermittent waters has had a role in the survival of the subspecies by providing a reproductive advantage over fall-spawning nonnative salmonids now common within their range.

Spawning occurs wherever optimum-size gravel (12–85 mm in diameter) and optimum water temperatures (5.5–15.5°C) are found. Given these conditions, spawning occurred nearly equally in streams under forest canopy and in open meadow stream habitats (Cope 1957b). A typical female Yellowstone cutthroat trout of 394 mm TL in Clear Creek will deposit about 1,300 eggs (2,633 eggs/kg) in gravel (Jones et al. 1985). Mills (1966) estimated Yellowstone cutthroat trout egg mortality in natural redds to range between 12 and 42%, mostly due to inadequate water flow in the substrate. Eggs typically hatch in 25–30 d (310 degree days, the sum of mean daily temperatures above 0°C), and juveniles emerge about 2 weeks later. Juveniles congregate in shallow, slow-moving parts of the stream, and those from migratory parents begin downstream (or upstream) movements shortly thereafter (Benson 1960; Ball and Cope 1961; Mills 1966).

Depending on variations in growth, spawning populations are comprised of individuals age 3 and older (primarily ages 4–7) (Irving 1955; Bulkley 1961; Jones et al. 1985). Age composition of a spawning population is susceptible to change due to angling exploitation (Gresswell and Varley 1988). In slow-growing, unfished populations, maximum life expectancy may extend to 11 years (Gresswell 1980).

In Yellowstone Lake, virtually all trout exceeding 300 mm TL are mature, and almost all fish below 250 mm are immature (Benson and Bulkley 1963). In alpine and subalpine lakes and streams, harsh conditions result in long-lived, slow-growing populations that mature at 100–130 mm TL.

Growth rates of Yellowstone cutthroat trout are highly variable depending on the stock and environmental conditions. Typical back-calculated lengths (TL) may be described as follows: age 1, 10 cm; age 2, 18 cm; age 3, 24 cm; age 4, 31 cm; age 5, 37 cm; age 6, 41 cm (Carlander 1969). Extensive growth and growth-related data for the subspecies were given by Irving (1955), Laakso (1956), Laakso and Cope (1956), Bulkley (1961), and Benson and Bulkley (1963).

Males appear to grow faster than females in Henrys Lake (Irving 1955) and Yellowstone Lake. Males commonly attain longer length, but this may be due to greater longevity rather than faster growth.

The largest individuals of the subspecies have been reported outside the historic range. An age-7 Yellowstone cutthroat trout of 730 mm TL and 6.8 kg was caught in Strawberry Reservoir, Utah (Platts 1958). This reservoir also yielded what may be the world record Yellowstone cutthroat trout, a 14.6 kg trout reportedly caught in 1948 (Roderick Stone, Utah Division of Wildlife Resources, personal communication).

Recreational Potential

Recreational potential of the Yellowstone cutthroat trout is extensive. Varley (1975) found that angling quality, as measured by landing rate, mean length, angler success (percent of single-day anglers landing one or more fish), and angler satisfaction, was higher for Yellowstone cutthroat trout than for any of the other seven sport-fish species found in Yellowstone National Park. We used the same parameters to produce a recreational value index based on creel survey data collected in 1986; Yellowstone cutthroat trout were rated above all other sport species considered (U.S. Fish and Wildlife Service, unpublished data). It is clear that high landing rates and angler success, coupled with relatively large mean length (363 mm), are important values to the majority of anglers.

High catchability of the Yellowstone cutthroat trout makes the subspecies an important and popular recreational resource; however, this vulnerability can easily lead to overharvest (Gresswell 1985). In Yellowstone Lake, annual angler effort of 15.8 h/hectare caused a significant decline in mean length and landing rate during the late 1960s (Gresswell and Varley 1988); the creel limit during that period was only three cutthroat trout of any size. In 1981, Schill et al. (1986) examined a 10.2 km section of the Yellowstone River (in Yellowstone National Park) between Fishing Bridge and Buffalo Ford (catch-and-release fishing only) and found that Yellowstone cutthroat trout were captured an average of 9.7 times during a 108-d season. Many cutthroat trout tagged during that study were captured two or three times in a single day. Hooking mortality was low; the mortality rate per single capture was 0.3%, and about 3% of the population died during the study as a result of catch-and-release by anglers (Schill et al. 1986).

Varley (1984) estimated the monetary value of a wild Yellowstone cutthroat trout fishery managed under catch-and-release-only restrictions. Using the actual price of cutthroat trout purchased from commercial hatcheries in Idaho during 1981, he estimated a cost of $1.55 per fish to replace the average Yellowstone cutthroat trout (391 mm) in the Yellowstone River. He reasoned that this value was for a cutthroat trout captured and removed, but released fish would maintain this value as an "avoided cost" of replacement. Based on the mean number of captures reported by Schill et al. (1986) and an average estimated catchable-life of 3 years, each cutthroat trout was worth $45 in avoided replacement costs. This illustrates the value of a wild Yellowstone cutthroat trout fishery and represents a cost that would exceed the ability of a public agency to support the fishery with hatchery trout.

In some situations, however, Yellowstone cutthroat trout have become an important component of hatchery programs. The Montana Department of Fish, Wildlife, and Parks has developed a brood stock of the Yellowstone subspecies originally derived from eggs obtained in 1969 from McBride Lake, Yellowstone National Park (Dean and Mills 1970). The state has used the McBride strain for establishing wild, reproducing populations of Yellowstone cutthroat trout. Additionally, this hatchery stock has been used to supplement wild stocks in areas where natural reproduction is limited (Dotson 1985). The McBride Lake stock of Yellowstone cutthroat trout has exhibited a long life span, and the fish spawn successfully if habitat is suitable. Random mating and brood-stock selection in conjunction with reintroduction of wild gametes every third generation have minimized the loss of genetic diversity in the hatchery brood stock (McMullen and Dotson 1988, this volume). Annual evaluations of enzyme polymorphisms by electrophoretic techniques have indicated little change from the wild stock.

The Wyoming Game and Fish Department also maintains a brood stock of Yellowstone cutthroat trout. Some problems have been experienced with the genetic purity of the brood stock, but a program has been initiated to evaluate the stock and replace it if necessary (John Baughman, Wyoming Game and Fish, personal communication). Managers in Wyoming have found that stocked Yellowstone cutthroat trout have returned to the creel at a rate twice that of hatchery rainbow trout (Kent 1984), a fact that has greatly increased the demand for a hatchery stock of the Yellowstone subspecies.

The Idaho Department of Fish and Game no longer maintains a hatchery brood stock for Yellowstone cutthroat trout, but eggs are collected annually from Henrys Lake (IDFG 1986). Resulting fry are generally planted back into the lake, but some have been used to reestablish Yellowstone cutthroat trout in areas where they are threatened.

Management

A rigorous examination of management practices used for Yellowstone cutthroat trout fisheries is beyond the scope of this paper; however, a brief discussion of current management trends follows. In general terms, management of the Yellowstone subspecies can be divided into (1) wild trout programs for native and introduced populations, and (2) maintenance stocking from hatchery brood stocks. Genetic integrity, habitat management, and special angling regulations are important elements which influence these programs.

Genetic Integrity

Within the past decade, management agencies have increased efforts to identify genetically unaltered populations of Yellowstone cutthroat trout. Primary emphasis has been on the detection of hybridization with rainbow trout (Loudenslager and Gall 1980, 1981) and other cutthroat trout subspecies (Wishard et al. 1980; Leary et al. 1987). Although the need for this type of work may seem intuitively obvious, the paucity of available information suggests that its importance cannot be overemphasized (Hadley 1984). In theory, all populations of Yellowstone cutthroat trout that seem to be genetically unaltered should be protected; however, management policies generally require that positive identification be made prior to the initiation of any action meant to reduce negative human impacts.

Maintenance of genetic integrity is a much more complicated process than it may seem initially. An attempt has been made in this paper to highlight the ecological diversity in one highly variable cutthroat trout subspecies. This is extremely difficult because so little is known about the entire range of ecological variation that is suspected to be present in the Yellowstone subspecies. Based on known diversity, however, stocks of the indigenous Yellowstone cutthroat trout display adaptations to different environments and biotic communities that have resulted in ecotypes displaying characters as variable as those commonly found between subspecies or even between species of trout (Bulkley 1963; Loudenslager and Gall 1981). As such, the traditional trinomial system of nomenclature partially fails as a management tool.

When a highly variable species is known only by a single name, the institutional tendency seems to be to recognize and manage for the variation manifest in one entity. Legal and administrative protection available for those population segments representing the remaining unnamed (and often unknown) variation is significantly lessened. This variation is subject to loss by well-known factors such as nonnative species introductions, exploitation, and pollution, but more likely hazards are genetic dilution or hybridization.

Managers interested in the survival and perpetuation of the remaining genetic variability in the Yellowstone cutthroat trout complex need to acquire a tool beyond the classic taxonomic approach. To date, it seems that evidence of life history and behavioral variability in these forms has advanced faster than knowledge of their genetic or phenotypic traits. The stock concept provides a means to incorporate a genetic perspective into management decisions (MacLean and Evans 1981). This concept has become a integral part of fisheries management in the Great Lakes and in most marine fisheries (Kutkuhn 1981; Larkin 1981). Although questions arise concerning the definition of the word stock (Booke 1981), MacLean and Evans (1981) argue that adoption of the concept is critical to preserving valuable fishery resources.

Behnke (1972) suggested the development of a "breeders handbook" to characterize the various biological attributes and document genetic diversity of polytypic species such as cutthroat trout. Poultry breeders and agronomists (among others) have extensively organized compendia to serve their particular industries. A registry of ecotypic varieties of cutthroat trout would help preserve genetic variation for its practical, scientific, and aesthetic value. It would also significantly advance the management of cutthroat trout as sport fishes.

Habitat Management

Habitat management comprises protection, enhancement, and improvement of aquatic environments. Protection involves preventing degradation of aquatic habitat from pristine condition. Enhancement and improvement have a common goal of increasing the productive potential of a stream habitat; however, enhancement is associated with rehabilitative efforts (Platts and Rinne 1985), and improvement may include habitats previously unaffected by human activities as well as those that have been so affected. Perpetuation of Yellowstone cutthroat trout depends on continued protection of unaltered habitats, but enhancement activities are urgently needed in many portions of the current and historical range of the subspecies.

Stream dewatering, due mostly to irrigation demands, is a major limiting factor for reproduction in many waters inhabited by Yellowstone cutthroat trout (Clancy 1988; Thurow et al. 1988, both this volume). This is not a new problem, however, and many western states are attempting to legally establish fish and wildlife sustenance as a "beneficial" use of flow waters. Of all possible enhancement activities, maintenance of adequate streamflow may be the most essential but most difficult to attain.

According to Platts and Rinne (1985), recognition of the importance of riparian habitats to fisheries management has developed primarily since the late 1970s. They noted that because riparian communities often respond quickly to management changes, these habitats can often be restored. The role of riparian vegetation has been discussed by Platts (1983) and Moring et al. (1985), and Platts and Rinne (1985) reviewed stream enhancement projects (including those involving riparian areas) throughout the Rocky Mountains.

Although stream enhancement may not be as critical, in a restricted sense, as adequate instream flow is to the future of the Yellowstone cutthroat trout, it has received increased emphasis by most management agencies in recent years. Riparian habitats have often been substantially improved by altering livestock grazing strategies (Platts and Rinne 1985). Because much of the current range of the subspecies lies within areas used for summer livestock grazing, innovative riparian management activities provide a unique opportunity to improve the status of the Yellowstone cutthroat trout.

Special Regulations

Special regulations have been defined as the use of number limits, size limits, and terminal gear specifications, either singly or in combination, to reduce sport-fishery harvest (Gresswell, in press). If angler harvest is a limiting factor for a particular fishery, special regulations can provide protection necessary to sustain fish populations under high levels of angler effort. Habitat degradation resulting from human influences is negligible in Yellowstone National Park, and fisheries within the area have provided an opportunity to evaluate the effects of angler harvest where maintenance stocking is prohibited and habitat is protected (Varley 1980; Gresswell and Varley 1988; Gresswell, in press). Yellowstone cutthroat trout have responded positively to catch-and-release-only (no-kill) regulations in all cases. A two-fish, 330-mm maximum-size limit provided similar results on Yellowstone Lake (Gresswell and Varley 1988), but a rapid decline in landing rate and mean length were noted in Riddle Lake under the same regulation. The apparent failure of the 330-mm maximum-size limit on Riddle Lake emphasizes the need to understand the relationship between population size structure, maturity, and catchability (Gresswell, in press).

Special regulations have gained popularity among anglers and fishery managers throughout the country in the past decade, and these regulations are an important element in the sustenance of Yellowstone cutthroat trout within their current range. Clancy (1987) and Thurow et al. (1988) discussed results of catch-and-release-only regulations for fluvial populations in Montana and Idaho. In Wyoming, special regulations have been considered for Yellowstone cutthroat trout in portions of the historic range of the subspecies (Kent 1984). Although angler harvest may not be the primary cause of a diminished fishery resource, special regulations should be incorporated into any program to restored native or introduced populations of Yellowstone cutthroat trout to their former abundance.

References

Albright, H. M. 1920. Superintendent's annual report for Yellowstone National Park 1919. U.S. National Park Service, Yellowstone National Park, Wyoming.

Ball, O. P. 1955. Some aspects of homing in cutthroat trout. Proceedings of the Utah Academy of Sciences, Arts, and Letters 32:75–80.

Ball, O. P., and O. B. Cope. 1961. Mortality studies on cutthroat trout in Yellowstone Lake. U.S. Fish and Wildlife Service Research Report 55.

Baxter, G. T., and J. R. Simon. 1970. Wyoming fisheries. Wyoming Game and Fish Commission Bulletin 4.

Behnke, R. J. 1972. The rationale of preserving genetic diversity. Proceedings of the Annual Conference Western Association of Game and Fish Commissioners 52:559–561.

Behnke, R. J. 1979. The native trouts of the genus *Salmo* of western North America. Report to the U.S. Fish and Wildlife Service, Denver, Colorado.

Behnke, R. J. 1988. Phylogeny and classification of cutthroat trout. American Fisheries Society Symposium 4:1–7.

Behnke, R. J., and M. Zarn. 1976. Biology and management of threatened and endangered western trout. U.S. Forest Service General Technical Report RM-28.

Benson, N. G. 1960. Factors influencing production of immature cutthroat trout in Arnica Creek, Yellowstone Park. Transactions of the American Fisheries Society 89:168–175.

Benson, N. G., and R. V. Bulkley. 1963. Equilibrium yield management of cutthroat trout in Yellowstone Lake. U.S. Fish and Wildlife Service Research Report 62.

Binns, N. A., and F. M. Eiserman. 1979. Quantification of fluvial trout habitat in Wyoming. Transactions of the American Fisheries Society 108:215–228.

Booke, H. E. 1981. The conundrum of the stock concept—are nature and nuture definable in fishery science? Canadian Journal of Fisheries and Aquatic Sciences 38:1479–1480.

Bowler, B. 1975. Factors influencing genetic control in lakeward migrations of cutthroat trout fry. Transactions of the American Fisheries Society 104:474–482.

Briggs, J. C. 1953. The behavior and reproduction of salmonid fishes in a small coastal stream. California Department of Fish and Game, Fish Bulletin 94.

Brown, C. J. D. 1971. Fishes of Montana. Big Sky Books, Montana State University, Bozeman.

Bulkley, R. V. 1961. Fluctuations in age composition and growth rate of cutthroat trout in Yellowstone Lake. U.S. Fish and Wildlife Service Research Report 54.

Bulkley, R. V. 1963. Natural variation in spotting, hyoid teeth counts, and coloration of Yellowstone cutthroat trout, *Salmo clarki lewisi*. U.S. Fish and Wildlife Service, Special Scientific Report—Fisheries 460.

Carlander, K. D. 1969. Handbook of freshwater fishery biology, volume 1, 3rd edition, Iowa State University Press, Ames.

Clancy, C. 1987. Inventory and survey of waters of the project area. Montana Department of Fish, Wildlife, and Parks, Federal Aid in Fish Restoration, Project F-9-R-35, Job 1-C, Job Progress Report, Helena.

Clancy, C. G. 1988. Effects of dewatering on spawning by Yellowstone cutthroat trout in tributaries to the Yellowstone River, Montana. American Fisheries Society Symposium 4:37–41.

Cope, O. B. 1956. Some migration patterns in cutthroat trout. Proceedings of the Utah Academy of Sciences, Arts, and Letters 33:113–118.

Cope, O. B. 1957a. Races of cutthroat trout in Yellowstone Lake. U.S. Fish and Wildlife Service Special Scientific Report—Fisheries 208:74–84.

Cope, O. B. 1957b. The choice of spawning sites by cutthroat trout. Proceedings of the Utah Academy of Sciences, Arts, and Letters 34:73–79.

Dean, J. L., and L. E. Mills. 1970. Fishery management program in Yellowstone National Park. U.S. Fish and Wildlife Service, Technical Report for 1969, Yellowstone National Park, Wyoming.

Dean, J. L., and J. D. Varley. 1974. Fishery management program in Yellowstone National Park. U.S. Fish and Wildlife Service, Technical Report for 1973, Yellowstone National Park, Wyoming.

Dotson, T. 1985. Broodstock management: part of the fisheries challenge. Montana Outdoors 162:34–37.

Erman, D. C., and V. M. Hawthorne. 1976. The quantitative importance of an intermittent stream in the spawning of rainbow trout. Transactions of the American Fisheries Society 105:675–681.

Fleener, G. C. 1952. Life history of the cutthroat trout *Salmo clarki* Richardson, in Logan River, Utah. Transactions of the American Fisheries Society 81: 235–248.

Fromm, R. J. Circa 1941. An open history of fish and fish planting in Yellowstone National Park. U.S. National Park Service, Yellowstone National Park, Wyoming.

Gresswell, R. E. 1980. Yellowstone Lake—a lesson in fishery management. Pages 143–147 *in* W. King, editor. Wild trout II. Trout Unlimited and Federation of Fly Fisherman, Vienna, Virginia.

Gresswell, R. E. 1985. Saving the dumb gene in Yellowstone: there is more to preservation than granola. Proceedings of the Annual Conference Western Association of Fish and Wildlife Agencies 65:218–223.

Gresswell, R. E. In press. Special regulations as a fishery management tool in Yellowstone National Park. *In* Proceedings of the conference on science in the national parks—1986. U.S. National Park Service, Fort Collins, Colorado.

Gresswell, R. E., and J. D. Varley. 1988. Effects of a century of human influence on the cutthroat trout of Yellowstone Lake. American Fisheries Society Symposium 4:45–52.

Hadley, K. 1984. Status report on the Yellowstone cutthroat trout (*Salmo clarki bouvieri*) in Montana. Montana Department of Fish, Wildlife, and Parks, Helena.

IDFG (Idaho Department of Fish and Game). 1986. Five year management plan. Idaho Fish and Game, Boise.

Irving, R. B. 1955. Ecology of the cutthroat trout in Henrys Lake, Idaho. Transactions of the American Fisheries Society 84:275–296.

Jahn, L. 1969. Movements and homing of cutthroat trout (*Salmo clarki*) from open-water areas of Yellowstone Lake. Journal of the Fisheries Research Board of Canada 26:1243–1261.

Jones, R. D., P. E. Bigelow, R. E. Gresswell, L. D. Lentsch, and R. A. Valdez. 1983. Fishery and aquatic management program in Yellowstone National Park. U.S. Fish and Wildlife Service, Technical Report for 1982, Yellowstone National Park, Wyoming.

Jones, R. D., P. E. Bigelow, R. E. Gresswell, and R. A. Valdez. 1982. Fishery and aquatic management program, Yellowstone National Park. U.S. Fish and Wildlife Service, Technical Report for 1981, Yellowstone National Park, Wyoming.

Jones, R. D., D. G. Carty, R. E. Gresswell, K. A. Gunther, L. D. Lentsch, and J. Mohrman. 1985. Fishery and aquatic management program in Yellowstone National Park. U.S. Fish and Wildlife Service, Technical Report for 1984, Yellowstone National Park, Wyoming.

Jones, R. D., D. G. Carty, R. E. Gresswell, C. J. Hudson, L. D. Lentsch, and D. L. Mahony. 1986. Fishery and aquatic management program in Yellowstone National Park. U.S. Fish and Wildlife Service, Technical Report for 1985, Yellowstone National Park, Wyoming.

Jones, R. D., R. E. Gresswell, K. A. Gunther, and L. D. Lentsch. 1984. Fishery and aquatic management program in Yellowstone National Park. U.S. Fish and Wildlife Service, Technical Report for 1983, Yellowstone National Park, Wyoming.

Jones, R. D., J. D. Varley, D. E. Jennings, S. M. Rubrecht, and R. E. Gresswell. 1979. Fishery and aquatic management program in Yellowstone National Park. U.S. Fish and Wildlife Service, Technical Report for 1978, Yellowstone National Park, Wyoming.

Jones, R. D., J. D. Varley, D. E. Jennings, S. M. Rubrecht, and R. E. Gresswell. 1980. Fishery and aquatic management program in Yellowstone National Park. U.S. Fish and Wildlife Service, Technical Report for 1979, Yellowstone National Park, Wyoming.

Jordan, D. S. 1891. A reconnaissance of the streams and lakes of the Yellowstone National Park, Wyoming in the interest of the United States Fish Commission. Bulletin of the U.S. Fish Commission 9:41–63.

Kent, R. 1984. Fisheries management investigations in the Upper Shoshone River drainage, 1978–82. Wyoming Game and Fish, Fish Division Completion Report, Cheyenne.

Kutkuhn, J. H. 1981. Stock definition as a necessary basis for cooperative management of Great Lakes fish resources. Canadian Journal of Fisheries and Aquatic Sciences 38:1476–1478.

Laakso, M. 1956. Body–scale regression in juvenile cutthroat from Yellowstone Lake. Proceedings of the Utah Academy of Sciences, Arts, and Letters 33:107–111.

Laakso, M., and O. B. Cope. 1956. Age determination in Yellowstone cutthroat trout by the scale method. Journal of Wildlife Management 20:139–153.

LaBar, G. W. 1971. Movement and homing of cutthroat trout (*Salmo clarki*) in Clear and Bridge creeks, Yellowstone National Park. Transactions of the American Fisheries Society 100:41–49.

Larkin, P. A. 1981. A perspective on population genetics and salmon management. Canadian Journal of Fisheries and Aquatic Sciences 38:1469–1475.

Leary, R. F., F. W. Allendorf, S. R. Phelps, and K. L. Knudsen. 1987. Genetic divergence and identification of seven cutthroat trout subspecies and rainbow trout. Transactions of the American Fisheries Society 116:580–589.

Liebelt, J. E. 1968. A serological study of cutthroat trout (*Salmo clarki*) from tributaries and the outlet of Yellowstone Lake. Master's thesis. Montana State University, Bozeman.

Loudenslager, E. J., and G. A. E. Gall. 1980. Geographic patterns of protein variation and subspeciation in cutthroat trout, *Salmo clarki*. Systematic Zoology 29:27–42.

Loudenslager, E. J., and G. A. E. Gall. 1981. Cutthroat trout a biochemical-genetic assessment of their status and systematics. University of California, Davis.

MacLean, J. A., and D. O. Evans. 1981. Assessing and managing man's impact on fish genetic resources. Canadian Journal of Fisheries and Aquatic Sciences 38:1889–1898.

McCleave, J. D. 1967. Homing and orientation of cutthroat trout (*Salmo clarki*) in Yellowstone Lake, with special reference to olfaction and vision. Journal of the Fisheries Research Board of Canada 24: 2011–2044.

McMullin, S. L., and T. Dotson. 1988. Use of McBride Lake strain Yellowstone cutthroat trout for lake and reservoir management in Montana. American Fisheries Society Symposium 4:000–000.

Miller, R. B. 1954. Comparative survival of wild and hatchery-reared cutthroat trout in a stream. Transactions of the American Fisheries Society 83:120–130.

Mills, L. E. 1966. Environmental factors and egg mortality of cutthroat trout (*Salmo clarki*) in three tributaries of Yellowstone Lake. Master's thesis. Colorado State University, Fort Collins.

Moring, J. R., G. C. Garman, and D. M. Mullen. 1985. The value of riparian zones for protecting aquatic systems: general concerns and recent studies in Maine. U.S. Forest Service General Technical Report RM-20:315–319.

Platts, W. S. 1958. Age and growth of the cutthroat trout in Strawberry Reservoir, Utah. Proceedings of the Utah Academy of Sciences, Arts, and Letters 35: 101–103.

Platts, W. S. 1959. Homing, movements, and mortality of wild cutthroat trout (*Salmo clarki* Richardson) spawned artificially. Progressive Fish Culturist 21: 36–39.

Platts, W. S. 1983. Vegetation requirements for fisheries habitats. U.S. Forest Service General Technical Report INT-157:184–188.

Platts, W. S., and J. N. Rinne. 1985. Riparian and stream enhancement management and research in the Rocky Mountains. North American Journal of Fisheries Management 5:115–125.

Raleigh, R. F., and D. W. Chapman. 1971. Genetic control in lakeward migration of cutthroat trout fry. Transactions of the American Fisheries Society 100:33–40.

Schill, D. J., J. S. Griffith, and R. E. Gresswell. 1986. Hooking mortality of cutthroat trout in a catch-and-release segment of the Yellowstone River, Yellowstone National Park. North American Journal of Fisheries Management 6:226–232.

Scott, W. B., and F. J. Crossman. 1973. Freshwater fishes of Canada. Fisheries Research Board of Canada Bulletin 184.

Sigler, W. F., and R. R. Miller. 1963. Fishes of Utah. Utah State Department of Fish and Game, Salt Lake City.

Simpson, J. C., and R. L. Wallace. 1978. Fishes of Idaho. The University Press of Idaho, Moscow.

Thurow, R. F., C. E. Corsi, and V. K. Moore. 1988. Status, ecology, and management of Yellowstone cutthroat trout in the Upper Snake River drainage, Idaho. American Fisheries Society Symposium 4: 25–36.

Varley, J. D. 1975. The Yellowstone fishery. Pages 91–96 *in* W. King, editor. Wild trout management symposium. Trout Unlimited, Vienna, Virginia.

Varley, J. D. 1979. Record of egg shipments from Yellowstone fishes, 1914–1955. U.S. National Park Service, Information Paper 36, Yellowstone National Park, Wyoming.

Varley, J. D. 1980. Catch-and-release fishing in Yellowstone Park. Pages 137–142 *in* W. King, editor. Wild trout II. Trout Unlimited and Federation of Fly Fisherman, Vienna, Virginia.

Varley, J. D. 1981. A history of fish stocking activities in Yellowstone National Park between 1881 and 1980. U.S. National Park Service, Information Paper 35, Yellowstone National Park, Wyoming.

Varley, J. D. 1984. The use of restrictive regulations in managing wild salmonids in Yellowstone National Park, with particular reference to cutthroat trout, *Salmo clarki*. Pages 145–156 *in* J. M. Walton and D. B. Houston, editors. Proceedings of the Olympic wild fish conference. Peninsula College, Fisheries Technology Program, Port Angeles, Washington.

Welsh, J. P. 1952. A population study of Yellowstone blackspotted trout (*Salmo clarki*). Doctoral dissertation. Stanford University, Stanford, California.

Wishard, L., W. Christensen, and P. Aebersold. 1980. Biochemical genetic analysis of four cutthroat trout tributaries to the Idaho Blackfoot Reservoir system. Idaho Fish and Game, Final Report, Boise.

American Fisheries Society Symposium 4:25–36,1988

Status, Ecology, and Management of Yellowstone Cutthroat Trout in the Upper Snake River Drainage, Idaho

RUSSELL F. THUROW

Idaho Department of Fish and Game, Box 2111, Hailey, Idaho 83333, USA

CHARLES E. CORSI

Idaho Department of Fish and Game, 1515 Lincoln Road, Idaho Falls, Idaho 83401, USA

VIRGIL K. MOORE

Idaho Department of Fish and Game, 600 South Walnut, Boise, Idaho 83707, USA

Abstract.—Distribution and abundance of wild Yellowstone cutthroat trout *Salmo clarki bouvieri* have declined considerably following human development of the Snake River Plain in Idaho. Viable wild populations are largely confined to headwater areas throughout their former range. Major indigenous populations remain in the Blackfoot, South Fork Snake, and Teton rivers and Willow Creek. Yellowstone cutthroat trout exist as nonmigratory and migratory races. Within the productive waters of the upper Snake River drainage, Yellowstone cutthroat trout may attain 8 or 9 years of age, lengths exceeding 600 mm, and weights of 3–4 kg. They are opportunistic feeders and become highly piscivorous where forage species are abundant. Spawning occurs from May through early July, and most fish mature at age 4 or 5. Repeat spawning occurs in alternate and consecutive years. Migratory populations generally remain in headwater rearing areas for 2 or 3 years, although some populations emigrate as fry. A combination of factors have contributed to the decline of genetically pure Yellowstone cutthroat trout in Idaho, including habitat degradation, genetic introgression, and exploitation. In recognition of the outstanding values of indigenous Yellowstone cutthroat trout, the Idaho Department of Fish and Game has adopted a policy to place "priority consideration" on the management of remaining wild stocks. Wild cutthroat trout populations are currently being managed through habitat restoration and protection, curtailment of species introductions, and the adoption of innovative regulations. Monitoring of Yellowstone cutthroat trout populations will continue in the future to enable their management as a discrete stock.

Within Idaho, the original cutthroat trout native to the Snake River system is the Yellowstone cutthroat trout *Salmo clarki bouvieri*. It is believed that Yellowstone cutthroat trout were replaced by rainbow trout *Salmo gairdneri* where they came in contact downstream from Shoshone Falls (Figure 1) and by westslope cutthroat trout *Salmo clarki lewisi* in the Salmon and Clearwater river drainages (Behnke 1979). Shoshone Falls isolated cutthroat trout from contact with rainbow trout, and the Yellowstone cutthroat trout remains the native trout in the upper Snake River basin. Hubbs and Miller (1948) also believed cutthroat trout were native to the Sinks drainages, a series of streams which historically were tributaries to the Snake River. Streams are presently isolated by the Snake River Plain, a massive lava flow that occurred several thousand years ago. Water from the Sinks streams flows underground for several hundred kilometers before entering the Snake River at Thousand Springs. Recent surveys indicate the presence of Yellowstone cutthroat trout in portions of the Sinks drainages, but it is unknown if they are native or the result of early hatchery stocking programs.

The upper Snake River basin features high-elevation drainages (exceeding 1,500 m) and a semiarid climate characterized by hot summers and cold winters. Summer daytime temperatures regularly exceed 32°C, and winter temperatures fall below −22°C. Annual precipitation averages 40–45 cm, and the majority occurs as snow. Peak stream discharges usually occur during a 2–6-week period beginning in April or May as a result of snowmelt. Timing and duration of runoff is influenced by elevation and snowpack. Flows decrease throughout the summer and increase with the onset of winter precipitation. Waters of the upper Snake River basin are generally productive, exhibiting high alkalinities (>150 mg $CaCO_3$/L) and large concentrations of various ions.

Indigenous fish fauna of the upper Snake River basin is represented by 4 families (Salmonidae, Cyprinidae, Catostomidae, and Cottidae), 7 gen-

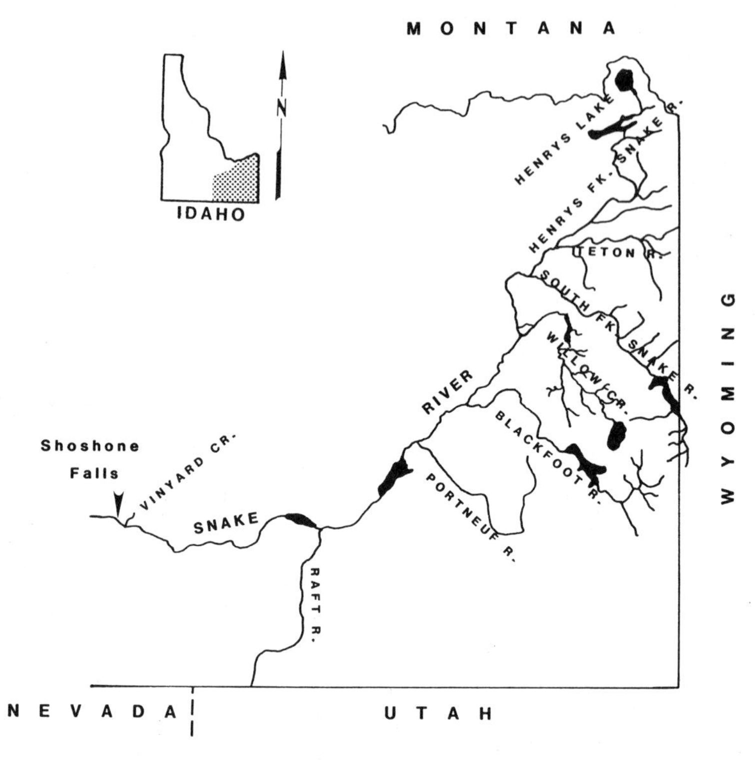

FIGURE 1.—Range of Yellowstone cutthroat trout in the Snake River above Shoshone Falls, Idaho. R = river; Cr. = creek; Fk. = fork.

era, and 11 species (Table 1). In the history of the upper Snake River basin, Yellowstone cutthroat trout have provided a popular and valuable fishery. As a result of human induced changes in the basin, distribution and abundance of this once outstanding resource have declined.

The intent of this paper is to present existing information on the status and ecology of the Yellowstone cutthroat trout in the Idaho portion of its native range. In addition, we explore those factors which have contributed to its decline, discuss current management strategies, and spec-

TABLE 1.—Fish present [+] or possibly present (?) in drainages of the upper Snake River basin. Species marked by an asterisk (*) are not indigenous to the basin.

		Distribution by drainage					
Common name	Species	Blackfoot River	Henrys Fork Snake River	South Fork Snake River	Teton River	Willow Creek	Portneuf River
Yellowstone cutthroat trout	*Salmo clarki bouvieri*	+	+	+	+	+	+
Brown trout*	*Salmo trutta*		+	+		+	+
Brook trout*	*Salvelinus fontinalis*	+	+		+	+	+
Rainbow trout*	*Salmo gairdneri*	+	+		+	+	+
Mountain whitefish	*Prosopium williamsoni*		+	+	+		
Longnose dace	*Rhinichthys cataractae*	+	+	+	+	+	+
Speckled dace	*Rhinichthys osculus*	+	+	+	+	+	+
Utah chub	*Gils atraria*	+	+	+	+	+	+
Common carp*	*Cyprinus carpio*	+				+	
Mountain sucker	*Catostomus platyrhynchus*	+	+	+	+	+	+
Utah sucker	*Catostomus ardens*	+	+	+	+	+	+
Bluehead sucker	*Catostomus discobolus*		?	+	?	+	?
Redside shiners	*Richardsonius balteatus*	+	+	+	+	+	+
Mottled sculpin	*Cottus bairdi*	+	+	+	+	+	+
Paiute sculpin	*Cottus beldingi*	+	+	+	+	+	+

ulate about the future of Yellowstone cutthroat trout in the upper Snake River drainage.

Methods

This paper is a synopsis of original work conducted by the authors and coworkers in the upper Snake River system. Methods of collecting fish included the use of electrofishing gear, trap nets, gill nets, and weirs. Creel surveys were used to obtain harvest information. Electrofishing gear included backpack units equipped with a hand-held anode, canoe or drift-boat mounted units with a hand-held or throwable anode, and jet-boat mounted units with hand-held or boom-mounted electrodes. All electrofishing in fluvial systems was conducted with direct current. Population estimates were completed either by removal methods (Seber and Le Cren 1967) or mark-and-recapture methods (Ricker 1975). Removal estimates were conducted in streams where sampling efficiency was sufficient to meet the requirements of the estimators. Mark-and-recapture estimates were conducted on larger streams.

Trap-netting and gillnetting were used to capture fish in lentic environments. Trap-netting was used to evaluate seasonal movements into and out of Blackfoot Reservoir. Trap-netting and gillnetting were used to assess species distribution in other upper Snake River system reservoirs.

Weirs were utilized to assess seasonal movements into and out of tributary streams (Everhart et al. 1975). Weirs were constructed to capture fish moving both upstream and downstream. Fry traps were also used to capture fry moving downstream. Fish traps were located near the mouths of streams, and we assumed that fish entering traps were moving into or out of a tributary.

Cutthroat trout movements were assessed from jaw-tagged or fin-clipped fish (Everhart et al. 1975) released at the point of capture. The use of fish traps enabled assessment of the timing of fish movements. Visual surveys of spawning grounds (Platts et al. 1983) enabled estimation of timing and number of spawners in tributaries to the Blackfoot River system.

Aging of cutthroat trout was accomplished by scale analysis, as described by Lagler (1956) and Royce (1972). Scales were removed from the area above the lateral line and posterior to the dorsal fin, mounted on slides, and viewed with a microprojector. Because cutthroat trout may not form an annulus their first year (Laakso and Cope 1956), scales were examined to determine if the number of circuli preceding the first annulus indicated retarded scale formation.

Angler surveys were of a stratified-random-sampling design (Malvestuto 1983) used to estimate harvest levels on various age-classes of fish. Creel checks provided information on the periods of vulnerability for various age groups and catch rates.

More detailed descriptions of methods and data analysis are found in Thurow (1982), Moore and Schill (1984), and Corsi (1988).

Life History

Yellowstone cutthroat trout in the upper Snake River drainage exist in both nonmigratory and migratory populations. Nonmigratory fish spend their entire life cycle in small tributary streams, and migratory fish utilize tributaries as spawning and juvenile rearing areas. Migratory populations may be either fluvial (all life stages occur in flowing waters) or adfluvial (a portion of the life cycle occurs in natural lakes or impoundments). In the Blackfoot River and Willow Creek systems, large impoundments have been formed in the lower reaches, and portions of fluvial populations have adapted to an adfluvial existence. Although resident and migratory populations occupying the same drainages function as discrete stocks, genetic interchange is probable.

Within the productive waters of the upper Snake River drainages, Yellowstone cutthroat trout have the potential to reach large sizes. Fish from the Blackfoot River and Willow Creek drainages have attained 8 or 9 years of age, lengths exceeding 500 mm, and weights of 2–3 kg. Occasional specimens exceed 600 mm total length (TL) and weigh more than 4 kg. Annual growth increments vary among subdrainages, and migratory stocks exhibit faster growth than nonmigratory fish (Table 2). Growth rates are largest for migratory fish following emigration from tributaries and prior to the onset of maturity. Mean size at scale formation was 47 mm with a range of 42–52 mm for emigrating fry from the South Fork Snake River tributaries. Most cutthroat trout populations examined formed an annulus during their first growing season, but retarded annulus formation was noted in colder tributaries of Willow Creek. Shorter growing seasons resulted in the annulus forming the following summer. Cutthroat trout from Yellowstone Lake tributaries were similar to those from Willow Creek both for fish length at time of squamation and for retarded

TABLE 2.—Calculated total lengths (mm) at each annulus formation for Yellowstone cutthroat trout from selected waters in the Snake River drainage, Idaho.

Drainage	Age (years)							Reference
	1	2	3	4	5	6	7	
Blackfoot River	117	213	321	403	442	473		Thurow (1982)
Henrys Lake	149	284	380	437	479	515		Irving (1954)
South Fork Snake River	86	184	277	343	410	450	480	Moore and Schill (1984)
Teton River basin	99	151	214	270	334			Irving (1979)
Willow Creek[a]	79	142	219	299	380	437		Corsi (1988)
Willow Creek[b]	81	139	198	242				Corsi (1988)

[a]Main-stem migratory stocks.
[b]Nonmigratory tributary stocks.

annulus formation (Brown and Bailey 1952; Laakso and Cope 1956).

A shift in the foraging pattern of migratory fish from a predominantly insectivorous pattern (Skinner 1985) to a largely piscivorous one following emigration is one reason why migratory fish grow larger than their resident counterparts. An exception occurs in Henrys Lake, Idaho, where fish feed on an abundant supply of large invertebrates (Irving 1954). It should be noted, however, that Yellowstone cutthroat trout are primarily opportunistic feeders and will consume food items of preferred size as they become available (McMasters 1970; Thurow 1982).

In Idaho, stream populations of Yellowstone cutthroat trout spawn from May through early July. The spawning run into Henrys Lake Hatchery begins in March and peaks during April and May, but this is an artifact of previous hatchery practices that selected early-spawning fish. Timing of tributary spawning in the Henrys Lake system is consistent with other wild populations. Following spawning, adult migratory cutthroat trout seldom remain in tributaries more than 2–4 weeks (Figure 2). Monthly population estimates of cutthroat trout in Diamond Creek showed adults present only in June (Neve and Moore 1983).

Females tend to outnumber males in spawning populations, with the notable exception of the Henrys Lake Hatchery run. In the Blackfoot River system, the female-to-male ratio ranged from 2.7:1 to 5.9:1 (Thurow 1982); and in the South Fork Snake River, the observed ratio was 1.2:1 (Moore and Schill 1984). Females are more likely to repeat spawn than males. Thurow (1982) found that 93% of repeat spawners were females. Repeat spawning may occur in alternative or consecutive years, with repeat spawners constituting up to 15% of some runs (Thurow 1982; Moore and Schill 1984).

In riverine portions of the upper Snake River drainage, most Yellowstone cutthroat trout mature at age 4 or 5, with some fish maturing at ages

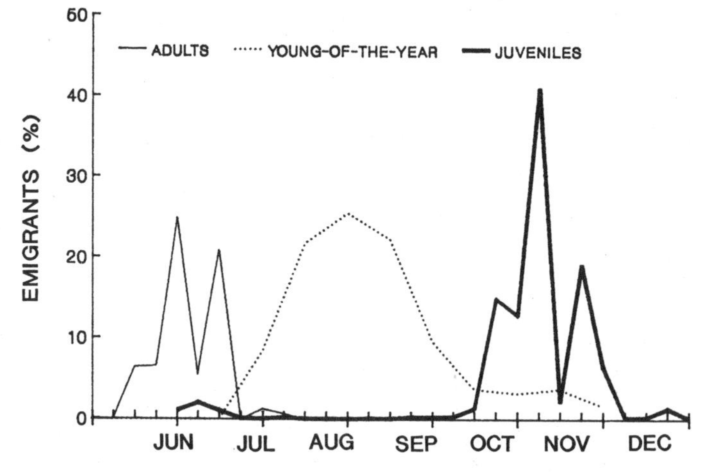

FIGURE 2.—Emigration timing for Yellowstone cutthroat trout young of the year, juveniles, and adults (spent spawners) from spawning tributaries of the upper Snake River drainage, Idaho.

3, 6 or 7. At Henrys Lake, where growth is more rapid, cutthroat trout typically mature at age 3. Mature fish less than 200 mm TL are seldom encountered and, in migratory populations, spawners usually exceed 275 mm and average between 300 and 500 mm (Irving 1954; Thurow 1982; Moore and Schill 1984; Corsi 1988). Fecundity is also variable with size (Platts 1958). Females from Henrys Lake averaging 319 mm, 408 mm, or 518 mm produced an average of 1,577, 1,914, or 2,930 eggs, respectively. The average female spawners (377 mm) from the South Fork Snake River produced 1,413 eggs (Moore and Schill 1984).

Fry emergence normally begins in mid-July and continues into fall, depending on the stream temperatures. In tributaries to Henrys Lake and in Burns Creek, a South Fork Snake River tributary, most juveniles emigrate as fry shortly after emergence (Figure 2; Irving 1954; Moore and Schill 1984). Thurow (1982) and Neve and Moore (1983) also observed cutthroat trout fry emigrating from intermittent Blackfoot River tributaries. In Henrys Lake and in a slough at the mouth of Burns Creek, ideal growth conditions exist for fry, and these stocks may have adapted to these conditions. Platts (1958) found most fry from Strawberry Reservoir tributaries in Utah (Henrys Lake cutthroat trout stock) emigrated at 35–70 mm over a 4–6-week period; similar behavior occurs in the fishes natural range in Idaho.

In other tributary systems, most juvenile Yellowstone cutthroat trout rear for 1–3 years prior to emigrating. Size at emigration is probably a function of the amount and quality of overwintering habitat (Chapman and Bjornn 1969). Within Willow Creek tributaries, beaver impoundments provide excellent winter habitat for yearling fish, and most emigrate after 1–2 years in the tributary (Corsi 1988). Emigrations of juvenile cutthroat trout are commonly associated with increased streamflow in spring and rapid declines in water temperature in the fall. Thurow (1982), Moore and Schill (1984), and Corsi (1988) all observed a bimodal curve of juvenile cutthroat trout migration, with movement peaking between June and early August, minimal movement in August and September, and a second peak of movement during low-flow periods between late September and November (Figure 2). Juvenile Yellowstone cutthroat trout exhibit density-dependent downstream movements as a function of reduced available habitat.

Status

Historically, Yellowstone cutthroat trout were abundant in all tributaries of the Snake River above Shoshone Falls. Viable wild populations of migratory fish remain in four major tributary systems: the Blackfoot, South Fork Snake, and Teton rivers and Willow Creek (Figure 1). Other drainages have wild populations that are largely confined to headwater areas due to isolation by irrigation, hydroelectric dams, and habitat degradation.

Blackfoot River

The Blackfoot River system supports a population of genetically pure, self-sustaining Yellowstone cutthroat trout of migratory and nonmigratory stocks. Tributaries to the Blackfoot River, Blackfoot Reservoir, and the Blackfoot River main stem are integral components in the life history and ecology of native cutthroat trout.

Wild cutthroat trout are the most abundant game fish in the Blackfoot River and its tributaries. Abundance surveys estimated 163 cutthroat trout/km in the main stem and up to 3,000 cutthroat trout/km in tributary rearing areas (Thurow 1982). The population is composed of predominantly age-1 and age-2 cutthroat trout, with fish attaining 8 years of age and 600 mm TL. Upper valley tributaries function as principle spawning and rearing areas. Approximately 140 km of stream are accessible in 25 tributaries. Spawner densities observed from 1978 to 1980 (mean, 13–15 spawners/km) were 20% of those observed in 1961 (mean, 74 spawners/km).

The Blackfoot River drainage upstream from Blackfoot Dam supports an important sport fishery, approaching 200,000 h of angler effort annually. Average lengths of cutthroat trout harvested from Blackfoot Reservoir and the Blackfoot River were 415 and 322 mm TL, respectively, and 58 and 28% exceeded 400 mm, respectively (Thurow 1982).

Angler harvest has increased cutthroat trout mortality. In 1959, mortality rates equaled 56% and 20% of the catch exceeded 500 mm TL (Thurow 1982). By 1978, mortality rates equaled 69% (Table 3) and only 4% of the catch exceeded 500 mm during the same period.

South Fork Snake River

The South Fork Snake River supports a migratory stock of Yellowstone cutthroat trout that utilizes tributaries for spawning and rearing areas.

TABLE 3.—Mortality coefficients (Z) and annual mortality rates (A) for Yellowstone cutthroat trout at various levels of angler exploitation.

Drainage	Angler effort		Z	A (%)	Source
	h/km	h/hectare			
Blackfoot River	303	216	1.18	69	Thurow (1982)
South fork Snake River, Idaho	758	85	1.78	83	Moore and Schill (1984)
South fork Snake River, Wyoming	482	54	1.33	73	Kiefling (1978)
Teton River					
Above dam	130	38	0.75	53	Moore and Schill (1984)
Below dam	183	53	1.17	69	Moore and Schill (1984)
Willow Creek	377	322	1.05	65	Corsi (1988)

Studies conducted during the late 1970s and early 1980s (Moore and Schill 1984) showed a declining cutthroat trout population, despite the fact they accounted for over 75% of the trout in the river.

Since 1984, the South Fork Snake River has been managed with special regulations applicable to 52 km of the 101 km of river below Palisades Dam. Regulations were specifically designed to reduce cutthroat trout harvest mortality by requiring release of all cutthroat trout 254–405 mm TL. By late 1986, cutthroat trout density had increased 79% from preregulation estimates to 142 fish/hectare. In the spring of 1987, a density estimate in a general-regulation reach (no length restriction) indicated only 20 fish/hectare.

Presently, only two of six tributaries are consistently producing large numbers of cutthroat trout emigrants to the South Fork Snake River. Other tributaries that are accessible and contain suitable habitat have limited spawning areas or have major irrigation diversions that effectively remove most emigrants before they reach the river. Viable populations of nonmigratory cutthroat trout are found in all the tributaries.

Teton River

The Teton River drainage supports self-sustaining nonmigratory and migratory Yellowstone cutthroat trout stocks. Wild Yellowstone cutthroat trout are the most widely distributed trout in several tributary and main-stem areas. Bitch and Teton creeks support important cutthroat trout spawning and rearing habitats. Four other tributaries have limited spawning use.

Angling effort on the Teton River is about 80,000 h, reflecting a 60% increase between 1975 and 1980 (IDFG 1986). Catch rates have declined by 48% during the same period. In 1974 and 1975, total annual mortality rates ranged from 62 to 69% (Irving 1979). Current status information is incomplete, although it indicates a declining cutthroat trout population. A comprehensive fisheries investigation was initiated in 1987.

Willow Creek

Yellowstone cutthroat trout are the most abundant and widely distributed game fish in the Willow Creek drainage system and they are represented by both nonmigratory and migratory stocks. Willow Creek, Grays Lake Outlet, Ririe Reservoir, and nearly 20 tributaries provide habitat for various life stages or migratory fish.

Recent evaluations of the Willow Creek system (Corsi 1988) have documented a declining cutthroat trout population. Areas that once supported large numbers of fish now have only remnant populations. Presently, only three tributaries produce a substantial number of emigrants. Areas exhibiting low angler exploitation and locations with quality habitat support large densities of cutthroat trout. Cutthroat trout densities in tributaries ranged from 0 to 38 fish/100 m^2 and in main-stem reaches from less than 1 to 8 fish/100 m^2. Tributary populations are comprised of 1- and 2-year old fish; however, main-stem reaches tend to have more subadult and adult fish, with fish up to 8 years of age.

Willow Creek and its tributaries upstream from Ririe Reservoir sustain approximately 50,000 h of angler effort annually. Ririe Reservoir annually supports 100,000 to 150,000 h of angler effort. Angler effort is heaviest in the areas that are easily accessible. Angler-induced mortality is high, and harvest accounts for nearly 90% of total annual mortality (65%) on catchable-size fish.

Remaining Drainages

Status evaluations have not been completed on all remaining waters. The existing data base suggests that remnant migratory wild Yellowstone cutthroat trout populations and isolated nonmigratory populations exist in the Blackfoot River downstream from Blackfoot Dam; the Henrys Fork Snake, Portneuf, and Raft river drainages; McCoy Creek and the Salt River (tributaries to Palisades Reservoir on the upper South Fork Snake River); and the main-stem Snake River.

Other waters that sustain wild cutthroat trout include Henrys Lake and its tributaries. Henrys Lake is a shallow, productive lake at the headwaters of the Henrys Fork Snake River, and the lake has historically supported a wild cutthroat trout fishery (Irving 1954). Since 1924, Henrys Lake Hatchery personnel have collected cutthroat trout eggs to restock Henrys Lake and supply fish for other Idaho waters (IDFG 1986). Henrys Lake currently supports an important sport fishery for wild and hatchery cutthroat trout, with 100,000 h of angler effort annually.

Factors Affecting Status

Several factors have contributed to the decline of genetically pure Yellowstone cutthroat trout in the upper Snake River basin of Idaho; these include habitat degradation, genetic introgression, and exploitation.

Habitat Degradation

Construction of dams and reservoirs has severely restricted the range of viable wild Yellowstone cutthroat trout populations. Dams have isolated migratory cutthroat trout from tributary spawning and rearing areas in the Blackfoot, Portneuf, South Fork Snake, Teton, Henrys Fork Snake, and main-stem Snake rivers. Lack of adequate spawning and rearing areas in reaches below dams has contributed to the decline of migratory cutthroat trout. For example, within the South Fork Snake River drainage, Palisades Dam terminated upstream movement of cutthroat trout to the majority of tributaries. A run of large (up to 2.3 kg) migratory cutthroat trout disappeared from the river 3–5 years after completion of the dam. Miller and Roby (1957) reported large numbers of migratory cutthroat trout congregating in the after-bay of the dam while attempting to reach upstream spawning areas.

Failure of the Teton Dam in 1976 severely degraded habitat in the Teton and Henrys Fork Snake rivers. Mass failures of water saturated soils occurred above the dam and severe channel scour occurred below the dam. An estimated 40–80% loss of cutthroat trout production occurred.

Dams and reservoirs have also indirectly contributed to the decline of native Yellowstone cutthroat trout. A majority of the reservoirs in the upper Snake River system are relatively shallow and eutrophic, with no temperature stratification. Substrates are predominantly silt and contain few vascular plants. Nongame species, including Utah chub, Utah sucker, and common carp, thrive in these impoundments. Attempts to eradicate nongame species from the Blackfoot River and Willow Creek drainages in the 1960s and 1970s eliminated thousands of cutthroat trout. Fortunately, the migratory behavior of the cutthroat trout prevented extermination.

Diversion of water for irrigation is very detrimental to cutthroat trout in the upper Snake River drainage. The greatest impacts result from channel dewatering, severe flow reductions, movements of trout into unscreened irrigation ditches, and degraded water quality. Low flows especially limit cutthroat trout populations in reaches of the Blackfoot, Henrys Fork Snake, Portneuf, Raft, Teton, and main-stem Snake rivers and Willow Creek. Large numbers of cutthroat trout enter unscreened ditches in the South Fork Snake River, Willow Creek, and tributaries to Henrys Lake. In 1980, 1,589 cutthroat trout between 50 and 540 mm TL were salvaged from 1,100 m of ditch on the upper Blackfoot River (Thurow 1982). Irrigation return water degrades the water quality of the Portneuf River and Vinyard Creek.

Other agricultural practices have adversely affected wild Yellowstone cutthroat trout. Intensive livestock grazing has contributed to the deterioration of riparian areas by stream bank sloughing, channel instability, erosion, and sedimentation. In the Blackfoot River and Willow Creek tributaries, stream reaches altered by livestock display unstable banks and predominantly silt substrate (Platts and Martin 1978). Altered reaches sustained fewer spawning cutthroat trout and had smaller densities of rearing juveniles (Thurow 1982; Corsi 1988). Livestock grazing impacts are widespread in the upper Snake River basin, particularly in the Blackfoot, Portneuf, South Fork Snake, and Teton rivers; Henrys Lake tributaries; and Willow Creek. Impacts occur on private lands as well as lands administered by the Bureau of Land Management, U.S. Forest Service, and the Idaho Department of Lands. Within the upper Blackfoot River and Willow Creek, extensive areas of riparian willow habitat have been eradicated with herbicides. Dry-land wheat farming has contributed to large sediment inputs, particularly in the Willow Creek drainage, which is considered one of the 20 worst agricultural erosion areas in the nation (Moeller 1981).

Development of mineral and energy resources also has the potential to adversely affect Yellowstone cutthroat trout. Within the Blackfoot River drainage, accelerated phosphate mining has increased sediment levels in streams (Platts and Martin 1978; Thurow 1982). Proposed small hydropower developments further threaten fish by reducing flows, blocking passages, and causing entrainment of fish and degradation of riparian areas.

Genetic Introgression and Species Interactions

Introductions of nonnative salmonids have threatened indigenous Yellowstone cutthroat trout through hybridization and displacement. Rainbow trout have been the most widely introduced species in the upper Snake River basin, and they commonly hybridize with cutthroat trout. As Behnke (1979) observed, cutthroat trout that have evolved in isolation from rainbow trout may lack isolating mechanisms that allow for coexistence. The most notable example of displacement occurred in the Henrys Fork Snake River, where native Yellowstone cutthroat trout are nearly extinct and most rainbow trout exhibit evidence of hybridization with cutthroat trout. Rainbow trout have also displaced Yellowstone cutthroat trout in reaches of the lower Blackfoot, Portneuf, and Teton rivers.

In the remaining drainages, naturally reproducing rainbow trout have not become well established and hybridization with cutthroat trout is uncommon. Electrophoretic analysis of cutthroat trout from four tributaries to the upper Blackfoot River displayed no evidence of hybridization with rainbow trout (Wishard et al. 1982). Lack of extensive hybridization in these drainages may be partially due to temporal separation of late-spring spawning cutthroat trout (May, June) and winter or early-spring spawning hatchery rainbow trout stocks.

Nonnative brook trout have been introduced to reaches of the upper Snake River basin. They are abundant in tributaries to Willow Creek, the Blackfoot, Henrys Fork Snake, and Teton rivers. We do not believe that brook trout have displaced or outcompeted native cutthroat trout. Fausch and White (1981) suggested that environmental factors, fishing mortality, and predation may favor one species in certain situations and thereby effect changes that are difficult to separate from those caused by interspecific competition. Griffith (1972) indicated that brook trout may not effectively displace cutthroat trout. Interactive segregation in which one species is more efficient in a habitat than another (Nilsson 1963) may be occurring in some sympatric populations of Yellowstone cutthroat trout and brook trout. Brook trout are commonly most prevalent in meadow and headwater reaches of streams with physical environments exhibiting low facing velocities, upwelling ground water, overhanging bank cover, and beaver impoundments.

Differential angling mortality has probably contributed to the replacement of cutthroat trout by brook trout. MacPhee (1966) observed that cutthroat trout were more than twice as likely to be caught by anglers as brook trout. Brook trout also mature at early ages (Jensen 1971), and they often reach maturity before being fully recruited to the fishery. Angling effort capable of overexploiting late-maturing cutthroat trout may have minimal impact on early-maturing brook trout.

The nonmigratory nature of brook trout may also contribute to their replacement of Yellowstone cutthroat trout. Because migratory populations of cutthroat trout require a range of habitats, isolation from, or degradation of habitats can adversely affect a cutthroat trout population and have no affect on the brook trout population.

Brown trout have also been introduced in the upper Snake River basin. Viable populations occur in the Henrys Fork Snake, South Fork Snake, and main-stem Snake river drainages and in the Willow Creek drainage. Exploitation appears to be the primary factor limiting cutthroat trout in sympatry with brown trout. In the South Fork Snake River, special regulations specific to cutthroat trout (slot limit) resulted in a 79% increase in densities of cutthroat trout and virtually no change in brown trout densities.

Exploitation

Cutthroat trout may be more vulnerable to angling than any other trout or char. Consequently, exploitation can be a significant cause for the decline in populations accessible to anglers. As angler effort increases, annual mortality rates on cutthroat trout that are fully recruited to the fishery (generally age 3) also increase (Table 3). At mortality rates exceeding 65%, the Yellowstone cutthroat trout population is expected to decline. As Ricker (1975) observed, high angler exploitation can result in compensatory decline in natural mortality. Within the South Fork Snake

River, angling mortality increased from 21% for age-3 and age-4 fish to 62% for age-4 through age-7 fish (Moore and Schill 1984). Conversely, natural mortality decreased from 31% for age-3 and age-4 fish to 22% for age-4–7 fish. Angler exploitation and selective angling for large, older Yellowstone cutthroat trout has contributed to the decline of native cutthroat trout in the upper Snake River basin.

Management

In recognition of the outstanding values of indigenous salmonids, the Idaho Department of Fish and Game (IDFG) has adopted a policy whereby, "native wild stocks of resident trout will receive priority consideration in all management decision involving resident fish" (IDFG 1986). Specific management goals have been adopted on a drainage-by-drainage basis. All drainages of the upper Snake River basin that support wild Yellowstone cutthroat trout have management direction to maintain wild populations.

In order to maintain and enhance wild cutthroat trout populations, management strategies have focused on those factors which limit populations. Research has helped identified factors that are limiting cutthroat trout populations in specific drainages. Where data are insufficient, as in the Teton River basin, comprehensive investigations are pending.

Habitat Restoration and Protection

The most significant habitat restoration efforts are focused on restoring fish passage over barriers, screening irrigation diversions, improving management of riparian areas, and regulating non–point source pollution.

Plans are pending to provide fish passage on the Buffalo River (tributary to Henrys Fork Snake River) and the Teton River by constructing fish ladders. As funding becomes available, additional sites may be modified. Federal mitigation funds are potential sources for future projects. Considerable effort is being directed to provide fish passage on small tributary streams where culverts and irrigation diversions limit fish movements. Installation of ladders and structures to pool water at culverts have restored access to several kilometers of spawning habitat in the South Fork Snake River for migratory Yellowstone cutthroat trout.

Screening of irrigation diversions has reduced losses of migratory cutthroat trout and increased recruitment in several drainages. Screens currently operate on the Blackfoot, Henrys Fork Snake, and South Fork Snake rivers, and additional screens are pending in the South Fork Snake River and Henrys Lake systems. Despite an Idaho law which requires screening of diversions to prevent fish losses, the cost of projects (range, $15,000–50,000 per diversion) limits their application. Current research examining alternatives to screening may help develop a more widely applicable solution.

Destruction of riparian areas and instream habitat by excessive livestock use remains one of the most prevalent problems in the upper Snake River basin. Feasible solutions include exclosures, alternative grazing schemes, and termination of grazing in highly critical reaches. Social, economic, and political factors have, unfortunately, slowed the implementation process.

The State of Idaho is currently developing water quality standards that may provide a mechanism to alleviate non–point source pollution impacts. Projects to control runoff from highly erodible cropland have been initiated in the Willow Creek drainage and are pending in the South Fork Snake and Teton rivers.

Protection of existing habitat is being attempted through application of state and federal laws that govern development activities. The most widely applied State of Idaho legislative actions include the Stream Protection Act, instream flow legislation, and water quality standards. The Stream Protection Act governs all activities (except irrigation) associated with surface waters. Unfortunately, violations of the act are frequent. Instream flow legislation enables acquisition of instream maintenance flows for fish, wildlife, and recreation values. Securing these flows remains a priority for all waters supporting native Yellowstone cutthroat trout.

Species Interactions

To reduce the threat of genetic introgression, introductions of hatchery-reared cutthroat trout and rainbow trout have been eliminated from most waters of the upper Snake River basin that support self-sustaining populations of native Yellowstone cutthroat trout. Where nonnative introductions persist, techniques are being tested to reduce introgression. In Blackfoot Reservoir, morpholine (Cooper and Hasler 1976) is being used to imprint introduced Bear Lake cutthroat trout to specific areas to reduce their opportunity for straying. Experimental production of sterile

TABLE 4.—Regulations applied to maintain wild Yellowstone cutthroat trout in the upper Snake River drainage, Idaho.

Drainage and specific location	Regulation type[a]
Blackfoot River	
Main stem and tributaries above Blackfoot Reservoir	July 1 opening; bag limit, three fish
Henrys lake	
Outlet	Sanctuary—closed
Tributaries	August 1 opening; bag limit, two fish
Lake	October 31 closure; bag limit, two fish
South fork Snake River	
Pine Creek (tributary)	July 1 opening; bag limit, six fish
Burns Creek (tributary)	September 1 opening; bag limit, six fish
Main stem from Heise to Palisades Dam	Bag limit, two cutthroat trout; all cutthroat trout 250–405 mm long must be released; barbless hooks required
Teton River	
Teton Creek (tributary)	July 1 opening; bag limit, six fish
Willow Creek	July 1 opening; bag limit, six fish

[a]General regulations are six trout, only two larger than 405 mm, May 26 to November 30.

rainbow–Yellowstone cutthroat hybrid trout is being tested at Henrys Lake to reduce introgression with Yellowstone cutthroat trout (Rohrer and Thorgaard 1986).

Where brook and brown trout are sympatric with cutthroat trout, these nonnative species are being managed to provide additional angling opportunities. In drainages where quality habitat remains and angling mortality is not excessive, Yellowstone cutthroat trout seem to flourish, even in reaches shared with brook and brown trout. As Fausch and White (1981) observed, the mechanisms of change in sympatric populations are unclear. However, because brook and brown trout are less susceptible to harvest than cutthroat trout, regulations allow more liberal harvest of brook and brown trout than cutthroat trout. Regulations allow a daily bag limit of 10 brook trout from streams. In some drainages, as the South Fork Snake River, cutthroat trout are protected with a species-specific regulation, and brown trout are managed with general harvest regulation.

Past attempts to chemically eradicate nongame species have been ineffective and have been detrimental to cutthroat trout. Although competition for food between trout and nongame fish is cited as a cause for decreased trout populations, such competition is unlikely (Marrin and Erman 1982). Fishery managers in Idaho are presently seeking biological controls for nongame species, at the same time they recognize their importance as forage for cutthroat trout.

Regulations

A variety of regulations have been implemented in Idaho to reduce angling mortality and to restore populations of Yellowstone cutthroat trout. Regulations have been designed to (1) increase recruitment, (2) protect spawners, and (3) increase the numbers of large trout. Typically, regulations include size restrictions, bag limits, and seasonal closure.

A fundamental management goal is to protect spawning-size fish. Regulations have been implemented in several spawning areas of the upper Snake River basin to protect cutthroat trout migrating to and within spawning areas. With a few exceptions, angling in spawning streams begins on July 1, more than 30 d after the general trout season (Table 4). Local variations depend on the timing of spawning runs.

Regulations to expand Yellowstone cutthroat trout populations and to increase the numbers of large fish have been implemented in the Blackfoot and South Fork Snake river drainages. Within the Blackfoot River, management goals were developed to (1) reduce the present mortality of age-5 and older spawners from 69 to 55%, (2) double the density of cutthroat trout spawners in tributaries, increasing them from 15 to 30 fish/km, and (3) increase the landing rate from 0.4 to 0.7 fish/h (IDFG 1986). To accomplish these goals, bag limits were reduced from six to three fish in 1985 (Table 4). Quantitative evaluations are underway. Preliminary data suggest that the wild cutthroat trout population has not responded to the three-trout bag limit (D. Schill, IDFG, personal communication). More restrictive regulations may be required to meet stated management goals. Within the South Fork Snake River, goals were developed to increase the landing rate to 0.7 fish/h and to increase the percent of cutthroat trout larger than 405 mm to 16%. In 1984, a "slot limit" was

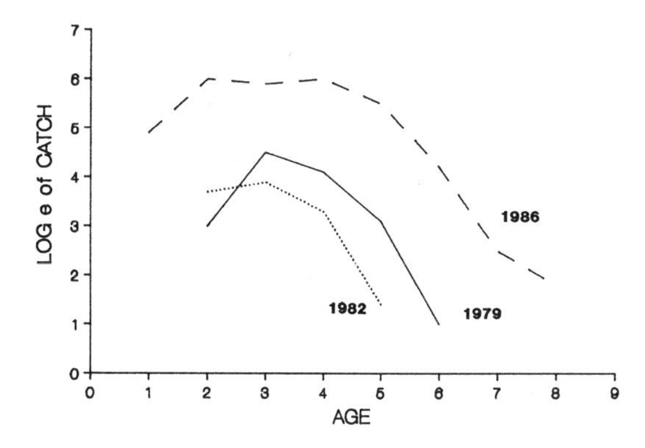

FIGURE 3.—Catch curves for Yellowstone cutthroat trout from the South Fork Snake River, Idaho.

implemented that allowed limited harvest of juvenile fish and mature fish that had spawned at least once (Table 4). After three seasons of special regulations, cutthroat trout densities increased by 79% to 142 fish/hectare. The percentage of cutthroat trout larger than 405 mm has increased from 1.9 to 9.1%. More cutthroat trout are attaining older ages, as reflected in the shift of the descending limb of the catch curve (Figure 3). Cutthroat trout and anglers are responding positively to the regulations. Due to a declining cutthroat trout population in the upper 12 km of the South Fork Snake River, the special regulation was expanded in 1988.

Conclusions

The future of native Yellowstone cutthroat trout in Idaho is promising. Longevity, rapid growth, and good catchability contribute to its sport value. After several decades of abuse, restoration efforts are finally being implemented. Emphasis will be focused on continued monitoring of native cutthroat trout populations to enable their management as discrete stocks.

In the future, protection of existing cutthroat trout habitat and restoration of degraded habitat will remain a top priority. Programs to reduce the threat of genetic introgression will also continue. Additional studies are needed to document the effect of brook and brown trout on cutthroat trout.

Finally, as additional data are collected, effects of angling mortality on cutthroat trout will be further evaluated. Where angler exploitation is excessive, regulations designed to reduce mortality rates will be tested. The relatively rapid responses of cutthroat trout special angling regulations, as in the South Fork Snake River, are cause for optimism in our efforts to restore this outstanding resource to its former abundance in the upper Snake River basin.

Acknowledgments

Regional fishery management personnel Bob Bell, Steve Elle, Scott Grunder, and Dan Schill provided data and reviewed the manuscript

References

Behnke, R. J. 1979. The native trouts of the genus *Salmo* of western North America. Report to U.S. Fish and Wildlife Service, Denver, Colorado.

Brown, C. J. D., and J. E. Bailey. 1952. Time and pattern of scale formation in Yellowstone cutthroat trout. Transactions of the American Fisheries Society 71:120–124.

Chapman, D. W., and T. C. Bjornn. 1969. Distribution of salmonids in streams with special reference to food and feeding. Pages 153–176 *in* T. G. Northcote, editor. Salmon and trout in streams. H. R. MacMillan Lectures in Fisheries, University of British Columbia, Vancouver.

Cooper, J. C., and A. D. Hasler. 1976. Electrophysiological studies of morpholine-imprinted coho salmon (*Oncorhynchus kisutch*) and rainbow trout (*Salmo gairdneri*) Journal of the Fisheries Research Board of Canada 33:688–694.

Corsi, C. E. 1988. The life history and status of the Yellowstone cutthroat trout (*Salmo clarki bouvieri*) in the Willow Creek drainage, Idaho. Master's thesis, Idaho State University, Pocatello.

Everhart, W. H., A. W. Eipper, and W. D. Youngs. 1975. Principles of fishery science. Cornell University Press, Ithaca, New York.

Fausch, K. D., and R. J. White. 1981. Competition between brook trout (*Salvelinus fontinalis*) and brown trout (*Salmo trutta*) for positions in a Michigan stream. Canadian Journal of Fisheries and Aquatic Sciences 38:1220–1227.

Griffith, J. S. 1972. Comparative behavior and habitat utilization of brook trout (*Salvelinus fontinalis*) and cutthroat trout (*Salmo clarki*) in small streams in northern Idaho. Journal of the Fisheries Research Board of Canada 29:265–273.

Hubbs, C. L., and R. Miller. 1948. Correlation between fish distribution and hydrographic history in the desert basins of the western United States. University of Utah Biological Series 10(7):17–166.

IDFG (Idaho Department of Fish and Game). 1986. Idaho fisheries management plan, 1986–1990. Boise.

Irving, J. S. 1979. The fish population in the Teton River prior to impoundment by Teton Dam with emphasis on cutthroat trout (*Salmo clarki*, Richardson). Master's thesis. University of Idaho, Moscow.

Irving, R. B. 1954. Ecology of the cutthroat trout in Henrys Lake, Idaho. Transactions of the American Fisheries Society 84:275–296.

Jensen, A. L. 1971. Response of brook trout (*Salvelinus*

fontinalis) populations to a fishery. Journal of the Fisheries Research Board of Canada 28:458–460.

Kiefling, J. W. 1978. Studies on the ecology of the Snake River cutthroat trout. Wyoming Game and Fish Department, Fisheries Technical Bulletin 3, Laramie.

Laakso, M., and O. B. Cope. 1956. Age determination in cutthroat trout by scale method. Journal of Wildlife Management 20:138–153.

Lagler, K. F. 1956. Freshwater fishery biology. Brown, Dubuque, Iowa.

MacPhee, C. 1966. Influence of differential angling mortality and stream gradient on fish abundance in a trout–sculpin biotope. Transactions of the American Fisheries Society 95:381–387.

Malvestuto, S. P. 1983. Sampling the recreational fisheries. Pages 397–419 *in* L. A. Nielson and D. L. Johnson, editors. Fisheries techniques. American Fisheries Society, Bethesda, Maryland.

Marrin, D. L., and D. C. Erman. 1982. Evidence against competition between trout and nongame fishes in Stampede Reservoir, California. North American Journal of Fisheries Management 2:262–269.

McMasters, M. J. 1970. The food habits of trout (*Salmonidae*) and sculpins (*Cottidae*) in two mountain streams. Master's thesis. Idaho State University, Pocatello.

Miller, T. W., and E. R. Roby. 1957. A progress report: South Fork Snake River, upper Snake River. U.S. Fish and Wildlife Service, Portland, Oregon.

Moeller, J. 1981. Water quality status report: Willow Creek report. Idaho Department of Health and Welfare, Division of Environment, WQ-47, Boise.

Moore, V., and D. Schill. 1984. South Fork Snake River fisheries inventory. River and stream investigations. Idaho Department of Fish and Game, Federal Aid in Fish Restoration, Project F-73-R-5, Job Completion Report, Boise.

Neve, L. C., and V. K. Moore. 1983. Population estimates and size regimes of cutthroat and brook trout in Diamond, Kendall, and Spring creeks, Idaho. Northwest Science 57:85–90.

Nilsson, N. 1963. Interactive segregation between fish species. Pages 295–313 *in* S. D. Gerking, editor. The biological basis of freshwater fish production. Blackwell Scientific Publication, Oxford, England.

Platts, W., W. Megahan, and G. Minshall. 1983. Methods for evaluating stream riparian and biotic conditions. U.S. Forest Service General Technical Report INT-138.

Platts, W. S. 1958. The natural reproduction of the cutthroat (*Salmo clarki*, Richardson) in Strawberry Reservoir, Utah. Master's thesis. Utah State University, Logan.

Platts, W. S., and S. B. Martin. 1978. Hydrochemical influences on the fishery within the phosphate mining of eastern Idaho. U.S. Forest Service Research Note INT-246.

Ricker, W. E. 1975. Computation and interpretation of biological statistics of fish populations. Fisheries Research Board of Canada Bulletin 191.

Rohrer, R. L., and G. H. Thorgaard. 1986. Evaluation of two hybrid trout strains in Henry's Lake, Idaho, and comments on the potential use of sterile triploid hybrids. North American Journal of Fisheries Management 6:367–371.

Royce, W. F. 1972. Introduction to the fishery sciences. Academic Press, New York.

Seber, G. A. F., and E. D. Le Cren. 1967. Estimating population parameters from catches large relative to the population. Journal of Animal Ecology 36: 631–643.

Skinner, W. D. 1985. Size selection of food by cutthroat trout (*Salmo clarki*) in an Idaho stream. Great Basin Naturalist 45:327–331.

Thurow, R. F. 1982. Blackfoot River fishery investigations. Idaho Department of Fish and Game, Federal Aid in Fish Restoration, Project F-73-R-3, Job Completion Report, Boise.

Wishard, L., W. Christensen, and P. Aebersold. 1982. Biochemical genetic analysis of four cutthroat tributaries to the Idaho Blackfoot Reservoir system. Final report to the Idaho Department of Fish and Game, Boise.

American Fisheries Society Symposium 4:37–41, 1988

Effects of Dewatering on Spawning by Yellowstone Cutthroat Trout in Tributaries to the Yellowstone River, Montana

CHRISTOPHER G. CLANCY

Montana Department of Fish, Wildlife and Parks, 406 South 11th Street Livingston, Montana 59047, USA

Abstract.—Dewatering of spawning tributaries caused by irrigation withdrawls appears to limit the recruitment of Yellowstone cutthroat trout *Salmo clarki bouvieri* into the Yellowstone River. Numbers of 2-year-old and older Yellowstone cutthroat trout were highest in sections of the river with high-quality tributaries and stable flows. Migration patterns of spawning fish suggest that Yellowstone cutthroat trout home to natal streams and that a small number of tributary streams provide most of the recruitment to a 133-km-long study reach of the river.

The Yellowstone cutthroat trout *Salmo clarki bouvieri* is classified a "species of special concern" within Montana by the Montana Department of Fish, Wildlife and Parks. It received this classification because this subspecies is found in only a small portion of its historic range within the state (Hanzel 1959; Hadley 1984). Presently, native fluvial populations are restricted to the Yellowstone River drainage, primarily upstream of Big Timber, Montana.

Yellowstone cutthroat trout in the Montana portion of the Yellowstone River spawn in tributary streams during late May through mid-July. Because large amounts of water are diverted from these tributaries during June through October for commercial cattle operations, many streams are dewatered during spawning and the early life stages of Yellowstone cutthroat trout.

The objectives of this study were to identify the effects of dewatering on populations and migration patterns of Yellowstone cutthroat trout in a section of the Yellowstone River.

Study Area

The Yellowstone River is one of the last free-flowing major rivers in the continental United States. It originates in northwestern Wyoming and flows through Yellowstone National Park before it enters Montana. In Montana, it flows in a northerly direction for 100 km to Livingston, where it turns to the east, and flows for 792 km to its confluence with the Missouri River in North Dakota. The primary land and water use in the area is agricultural, mostly commercial cattle operations (Anderson 1981).

Our study area was a 133-km segment of river between Yellowstone National Park and Springdale, Montana. It has a cobble–gravel stream bed and a moderate gradient of 2.1 m/km. The mean annual flow at the U.S. Geological Survey (USGS) gage near Livingston for 59 years of records between 1897 and 1984 was 107 m^3/s. Mean monthly flows ranged from 33 m^3/s for January to 381 m^3/s for June. Populations of brown trout *Salmo trutta*, rainbow trout *S. gairdneri*, Yellowstone cutthroat trout, and mountain whitefish *Prosopium williamsoni* inhabit the river. Nonsalmonid resident species include white sucker *Catostomus commersoni*, longnose sucker *C. catostomus*, mottled sculpin *Cottus bairdi*, and longnose dace *Rhinichthys cataractae*. The river has been managed as a wild trout fishery since 1972, when the last stocking of rainbow trout occurred.

Methods

Four study sections (Figure 1), each approximately 8 km in length, were electrofished annually during April and May from a 5.5-m-long aluminum boat powered by a 90-horsepower outboard jet unit. The boat was equipped with a double-boom electrode system based on designs by Novotny and Priegel (1974). Data collection began in 1978 and effort was intensified between 1982 and 1987.

Mark-and-recapture population estimates were calculated with Chapman's modification of the Petersen formula (Ricker 1975). One to three mark-and-recapture runs were required per estimate. Seven to 14 d separated the mark and recapture dates. Various river sections were electrofished for the purpose of tagging fish with individually numbered Floy FD-68B anchor tags to monitor seasonal movements.

Tributary steams were sampled for spawning activity by means of steel-frame traps in a few instances; however, most streams were electro-

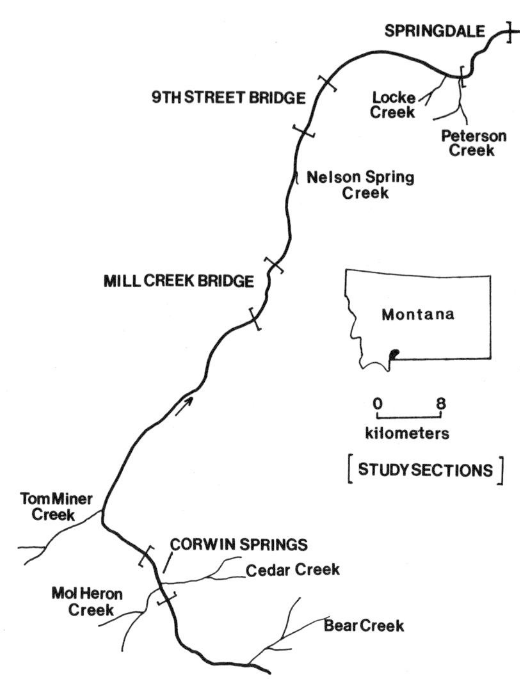

FIGURE 1.—Map of the upper Yellowstone River, showing the four study sections and the high-quality spawning tributaries.

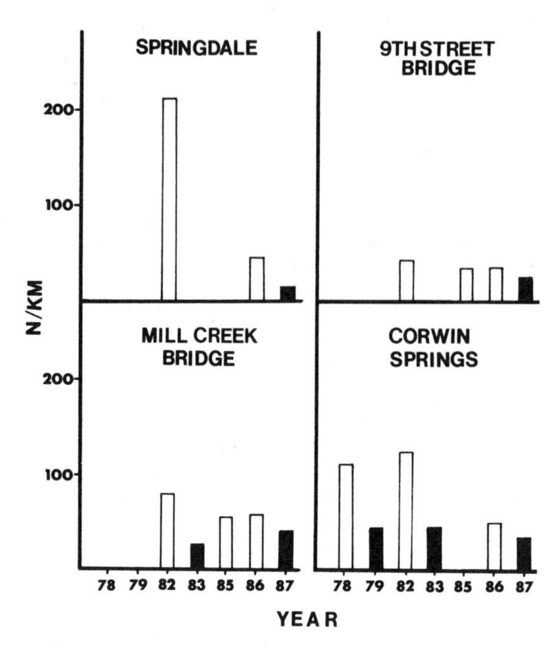

FIGURE 2.—Densities of 2-year-old Yellowstone cutthroat trout in the four study sections of the Yellowstone River. Black bars denote year classes spawned during low-water years. Only statistically valid estimates are included.

fished with a bank or a backpack unit, depending on the size of the stream. The lengths of study sections in tributary streams varied, so all comparisons were based on a standard section length of 305 m.

Because inconsistency of annulus formation is common with cutthroat trout (Lentsch and Griffith 1986), scales were not considered reliable for aging. Length-frequency distributions were used to separate 2-year-old fish from the older age-groups.

Flow information was obtained for two tributaries, the Shields River and Big Creek, where USGS gages are maintained. Although flows were not available for the remaining tributaries, a continuous-recording gage station at Carter's Bridge near Livingston provided an estimate of annual discharge in the Yellowstone River. The mean September flow in the Yellowstone River provided an index of the magnitude of the tributary flow during incubation and emergence periods.

Results

Populations

Population estimates of 2-year-old Yellowstone cutthroat trout in the four study sections were obtained during most years of the study (Figure 2). This age-group was used as an indicator of recruitment. The relationship between the numbers of 2-year-old and numbers of adult cutthroat trout is evident when the 1980 year class is considered. This year class was exceptionally strong as 2-year-olds in 1982 (Figure 2). During 1983 and 1984, elevated numbers of adult cutthroat trout (≥30 cm long) reflect this strong year class (Figure 3). By 1985 and 1986, the 1980 year class was losing dominance.

Mean September flows at the three gage sites were lowest in 1977, 1981, and 1985 (Table 1). Year classes of Yellowstone cutthroat trout that emerged during low-flow years tended to be weaker than year classes that emerged during years of higher flow (Figure 2). The variation of recruitment between years was least in the Ninth Street Bridge study section. Spawning in this section of river occurs in spring-fed creeks, primarily Nelson Spring Creek, whereas mountain snowpack is the source of water for the other tributary streams.

A stream was considered a high-quality spawning tributary if at least 10 spawners were annually captured per standard sample section (305 m) of stream during the peak of spawning. Tom Miner,

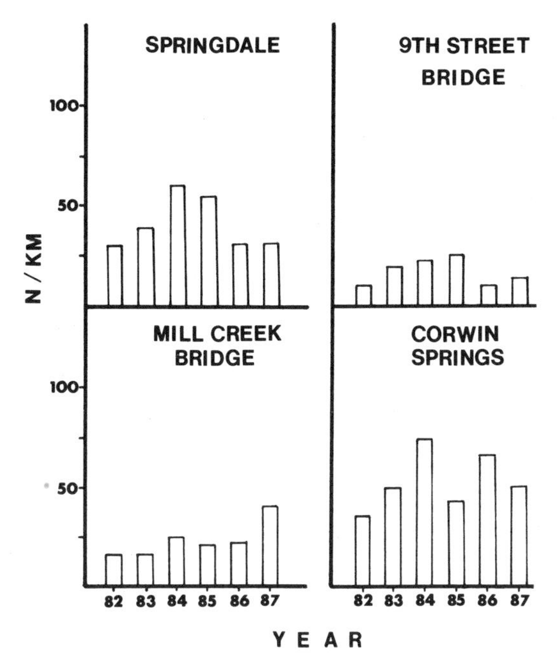

FIGURE 3.—Densities of adult Yellowstone cutthroat trout 30 cm long or longer in the four study sections of the Yellowstone River, 1982–1987.

Mol Heron, Cedar, and Bear creeks, near Corwin Springs; Nelson Spring Creek, near 9th Street Bridge; and Locke and Peterson creeks, near Springdale, were designated as high-quality spawning tributaries. Remaining streams do not support important runs because they are severely dewatered by irrigation withdrawls during midsummer.

The distribution of high-quality spawning tributaries appears to relate directly to the level of recruitment into the Yellowstone River study sections. Valid population estimates of 2-year-old Yellowstone cutthroat trout were collected for all four study sections in 1982, 1986, and 1987. Comparison of these estimates shows that the Springdale and Corwin Springs study sections generally supported higher numbers of 2-year-olds than did the 9th Street Bridge and Mill Creek Bridge study sections (Table 2). Study sections with the highest numbers of 2-year-olds were near the high-quality spawning streams.

TABLE 1.—Mean September flows (m^3/s) at the U.S. Geological Survey gage sites at Carter's Bridge on the Yellowstone River and on Shields River and Big Creek during the years indicated. Years marked with an asterisk (*) are considered low runoff years.

Year	Yellowstone River	Shields River	Big Creek
1976	85.0		1.0
1977*	51.7		0.7
1980	76.9	5.6	
1981*	50.3	2.2	
1983	76.7	5.2	0.9
1984	82.2	5.1	1.1
1985*	61.9	3.5	0.8

TABLE 2.—Mean numbers of 2-year-old Yellowstone cutthroat trout per kilometer of river during 1982, 1986, and 1987, and mean numbers of 3-year-old and older Yellowstone cutthroat during 1982–1987 in the four study sections of the Yellowstone River.

Study section	2-year-olds	3-year-olds and older
Springdale	97	48
9th Street Bridge	40	27
Mill Creek Bridge	63	34
Corwin Springs	75	106

The Mill Creek Bridge section supported unexpectedly high numbers of 2-year-old Yellowstone cutthroat trout despite the absence of high-quality tributaries nearby. Tag-return data suggest that the upper-basin tributaries in the area of Corwin Springs furnish recruits to this section (Table 3).

The Springdale and Corwin Springs study sections also supported the highest numbers of 3-year-old and older Yellowstone cutthroat trout (Table 2). The relatively high number of adults in the Mill Creek Bridge section probably reflected the results of catch-and-release regulations, which have allowed this population to increase in the past few years.

Migration

Adult cutthroat trout in the middle reaches of the study area, where there are few spawning tributaries, migrated long distances to spawn. Conversely, cutthroat trout in the upper and lower sections migrated shorter distances (Table 3). Yellowstone cutthroat trout from Springdale and Corwin Springs migrated from 1 to 8 km to a spawning stream, but the fish from the Mill Creek

TABLE 3.—Migration patterns of mature Yellowstone cutthroat trout in the upper Yellowstone River, Montana, 1983–1987.

Tagging location	Spawning stream (number of returns)
Springdale	Peterson (54), Locke (16), Nelson Spring (4)
9th Street Bridge	Nelson Spring (4)
Mill Creek Bridge	Tom Miner (7), Mol Heron (6)
Corwin Springs	Tom Miner (5), Mol Heron (18), Cedar (27)

TABLE 4.—Recaptures (locations and numbers) of spawning Yellowstone cutthroat trout tagged in tributary streams of the Yellowstone River

Recapture location	Original tagging location	
	Same	Different
Peterson Creek	9(100%)	0
Locke Creek	2(100%)	0
Nelson Spring Creek	5(100%)	0
Tom Miner Creek	40(100%)	0
Mol Heron Creek	14(93%)	1 (Peterson Creek)
Cedar Creek	5(100%)	0
Total	75(99%)	1(1%)

Bridge and 9th Street Bridge areas migrated 11–53 km (Figure 1).

Tag returns from repeat spawners during 1983–1987 suggest that mixing of fish between spawning tributaries is minimal (Table 4). Individual fish appeared to spawn in only one tributary, probably their natal stream (Berg 1975).

Discussion

The loss of spawning areas has been identified as a major reason for the decline of cutthroat trout populations. In this volume, Dwyer and Rosenlund (1988), Gerstung (1988), and Nielson and Lentsch (1988) report that diversion of water from spawning tributaries by agricultural users is one reason for the declines of the greenback cutthroat trout *S. c. stomias* in Colorado, the Lahontan cutthroat trout *S. c. henshawi* in California, and the Bear Lake population of the bonneville cutthroat trout *S. c. utah* in Utah, respectively. In Montana, the Yellowstone cutthroat trout population of the Yellowstone River also appears to be limited by the diversion of water from its spawning tributaries.

During low-water years, observation has shown that, under some circumstances, adults either do not spawn or interrupt their spawning to leave the stream as flows become inadequate. At other times, fish appear to spawn successfully but flows then drop to levels that expose the redds to desiccation. Variations in mountain snowpack, air temperature, and rainfall patterns also affect the success of spawning.

The particular stage in the life cycle of the Yellowstone cutthroat trout that is most sensitive to low flows is not well documented. The late summer months comprise the period in which cutthroat trout are incubating, emerging from the redds, and progressing through the early stages of their first year of life (Roberts and White 1986). Flows during this time appear to be very important to reproduction and recruitment. Becker et al. (1983) found that alevins of chinook salmon *Oncorhynchus tshawytscha* were more sensitive to dewatering than the eggs, indicating that reduced flows toward the end of the incubation period are more detrimental than earlier in the period. Hobbs (1937) and Reiser and White (1983) found that salmonid eggs that were kept moist, but were not in stagnant water, had a high survival rate. In most cases, dewatering in the Yellowstone River tributaries exceeds 8 weeks, and whether the eggs remain moist after the stream is dewatered is unknown. Berg (1975) conducted a survey of Yellowstone River tributaries during the peak of the irrigation seasons in 1974 and 1975 and found that 93 and 55 km, respectively, of streams were either dry or severely dewatered. He suggested that the complete diversion of most major tributaries of the upper Yellowstone River for irrigation was a significant factor in the decline of the Yellowstone cutthroat trout.

The presence of spring-fed creeks in the vicinity of the 9th Street Bridge study section allows for the stable recruitment. Because the flow in spring-fed creeks is fairly constant from year to year, and the creeks are not severely dewatered by irrigation, the production of recruits appears to be more stable than that occurring in other streams in the drainage.

The population of adult Yellowstone cutthroat trout is directly dependent upon recruitment of younger fish. Decreased recruitment caused by dewatering of spawning tributaries results in lower numbers of adult Yellowstone cutthroat trout.

Uneven distribution of spawning tributaries to the Yellowstone River also influences migration patterns of Yellowstone cutthroat trout. Because the majority of spawning occurred in the upper and lower portions of the study area, Yellowstone cutthroat trout from the middle sections migrated long distances to spawn in their natal streams. Tag returns suggested that the upper basin, which contains most of the spawning streams, was responsible for most of the recruitment to the middle section of the study area.

Solutions to the dewatering problem are presently not available. Although the Montana Department of Fish, Wildlife and Parks was granted instream flow reservations for these streams in 1978, existing water rights have historically diverted all of the midsummer water for irrigation purposes. Legal means for acquiring this water need to be pursued.

Dewatering of spawning tributaries contributes to the need to restrict angler harvest. It is clear that the reaches with the greatest number of high-quality spawning streams support the highest populations of Yellowstone cutthroat trout. Reduced recruitment from spawning tributaries has led to reduced numbers of adult fish and, thus, the capacity of the population to sustain angler harvest. Thurow et al. (1988, this volume) reports that the best fishing for cutthroat trout in the South Fork of the Snake River is found within 20 km of the tributaries that produce the most recruits. Catch-and-release regulations that were implemented in 1984 on the upper 80 km of the Yellowstone River in Montana have had some success in increasing the number of 30-cm-long and longer Yellowstone cutthroat trout (Clancy 1987). Although it is clear that acquisition of midsummer flows in the spawning tributaries is desirable, greater restriction of angler harvest is the only method available at this time to increase the population of Yellowstone cutthroat trout.

References

Anderson, B. 1981. How the river runs. A study of potential changes in the Yellowstone River basin. Montana Department of Natural Resources and Conservation, Final Report, Helena.

Becker, C. D., D. A. Neitzel, and C. S. Abernethy. 1983. Effects of dewatering on chinook salmon redds: tolerance of four development phases to one-time dewatering. North American Journal of Fisheries Management 3:373–382.

Berg, R. K. 1975. Fish and game planning, upper Yellowstone and Shields river drainages. Montana Department of Fish and Game, Federal Aid in Fish Restoration, Project FW-3-R, Job 1-C, Helena.

Clancy, C. G. 1987. Inventory and survey of waters of the project area. Montana Department of Fish, Wildlife and Parks, Federal Aid in Fish Restoration, Project F-9-R-35, Job Completion Report, Helena.

Dwyer, W. P., and B. D. Rosenlund. 1988. Role of fish culture in the reestablishment of greenback cutthroat trout. American Fisheries Society Symposium 4:75–80.

Gerstung, E. R. 1988. Status, life history, and management of the Lahontan cutthroat trout. American Fisheries Society Symposium 4:93–106.

Hadley, K. 1984. Status report on the Yellowstone cutthroat trout *Salmo clarki bouvieri* in Montana. Montana Department of Fish, Wildlife, and Parks, Helena.

Hanzel, D. A. 1959. The distribution of cutthroat trout *Salmo clarki* in Montana. Proceedings of the Montana Academy of Sciences 19:32–71.

Hobbs, D. F. 1937. Natural reproduction of quinnant salmon, brown and rainbow trout in certain New Zealand waters. New Zealand Marine Department, Fisheries Bulletin 6, Wellington.

Lentsch, L. D., and J. S. Griffith. 1986. Lack of first-year annuli on scales: frequency of occurrence and predictability in trout of the western United States, Pages 177–188 *in* R. C. Summerfelt and G. E. Hall, editors. Age and growth of fish. Iowa State University Press, Ames.

Nielson, B. R., and L. Lentsch. 1988. Bonneville cutthroat trout in Bear Lake: status and management. American Fisheries Society Symposium 4:128–133.

Novotny, D. W., and G. R. Priegel. 1974. Electrofishing boats; improved designs and operational guidelines to increase the effectiveness of boom shockers. Wisconsin Department of Natural Resources Technical Bulletin 73.

Reiser, D. W., and R. G. White. 1983. Effects of complete redd dewatering on salmonid egg-hatching success and development of juveniles. Transactions of the American Fisheries Society 112:532–540.

Ricker, W. E. 1975. Computation and interpretation of biological statistics of fish populations. Fisheries Research Board of Canada Bulletin 191.

Roberts, B. C, and R. G. White. 1986. Potential influence of recreational use on Nelson Spring Creek. Montana Cooperative Fishery Research Unit, Montana State University, Bozeman.

Thurow, R. F., C. E. Corsi, and V. K. Moore. 1988. Status, ecology, and management of Yellowstone cutthroat trout in the upper Snake River drainage, Idaho. American Fisheries Society Symposium 4: 25–36.

American Fisheries Society Symposium 4:42–44, 1988

Use of McBride Lake Strain Yellowstone Cutthroat Trout for Lake and Reservoir Management in Montana

STEVE L. MCMULLIN AND THURSTON DOTSON

Montana Department of Fish, Wildlife, and Parks, 1420 East Sixth Avenue Helena, Montana 59620, USA

Abstract.—Poor performance in Montana alpine lakes by hatchery-reared Yellowstone cutthroat trout *Salmo clarki bouvieri* was blamed on lack of adaptability and loss of genetic variability in the hatchery brood stock. A new brood stock was developed from wild gametes taken at McBride Lake in Yellowstone National Park. Mating and selection of fish to be used for future brood stock were done randomly to maintain genetic diversity as close as possible to that of the wild population. Genetic variability of brood stock was tested annually and wild gametes were reintroduced into the brood stock every third generation. Naturally reproducing populations of McBride strain Yellowstone cutthroat trout have been established in lakes with suitable inlet or outlet spawning habitats, usually after a single introduction. Performance in alpine lakes is characterized by rapid growth, high catchability, and a longevity of up to 9 years. McBride strain cutthroat trout are opportunistic feeders, adaptable to a variety of situations. Tests of the strain's ability to coexist with other species have not been completed. The keys to successful implementation of a strain management program are the selection of a strain that is well adapted to its intended use and the preservation of the strain's adaptability through maintenance of genetic variability.

Poor performance by hatchery fish in specific situations in the wild has led many fisheries managers to search for species or strains of fish naturally adapted to those situations. In some cases, poor performance in the wild may be due to domestication of, and loss of genetic variability among, the hatchery fish. Significant reduction in brood-stock genetic variability of westslope cutthroat trout *Salmo clarki lewisi* in Montana was noted within 14 years of their removal from the wild (Allendorf and Phelps 1980). When a domesticated brood stock is maintained for a long period of time, artificial selection for traits that make the fish easier to rear in the hatchery (but possibly less suited to survival in the wild) is likely to occur (Doyle 1983). These considerations led Montana fisheries managers to establish a new brood stock of Yellowstone cutthroat trout *Salmo clarki bouvieri* primarily for use in alpine lakes of south-central Montana.

Montana's Beartooth Plateau region, located northeast of Yellowstone National Park, provides many opportunities for fishing alpine lakes. Several hundred lakes, at elevations as high as 3,300 m, support fishable trout populations in or near the Absaroka–Beartooth Wilderness Area. The Montana Department of Fish, Wildlife, and Parks (MDFWP) began a comprehensive inventory of Beartooth Plateau lakes in the 1960s for the purpose of developing management plans. Most of the lakes were originally barren of fish. Various species have been introduced, including rainbow trout *Salmo gairdneri*, brown trout *Salmo trutta*, brook trout *Salvelinus fontinalis*, lake trout *Salvelinus namaycush*, and Arctic grayling *Thymallus arcticus*. The majority of the lakes, however, are managed for Yellowstone cutthroat trout, the only trout native to the area.

The MDFWP brood stock maintained for stocking these lakes originated from Yellowstone Lake. When egg-taking operations at Yellowstone Lake ceased in 1956, the source of wild gametes for the MDFWP brood stock was lost (Gresswell and Varley 1988, this volume). When the Beartooth Plateau lake inventory was initiated, it quickly became apparent that hatchery-reared Yellowstone cutthroat trout were not performing up to expectation in many alpine lakes. Survival and growth of hatchery fish was poor, and reproduction in the wild rarely occurred. Loss of genetic variability was suspected to be the major problem. However, an innate lack of ability to adapt to various conditions may also have been a factor. Behnke (1979) suggested that evolution of the Yellowstone Lake cutthroat trout in a stable environment, free of competition from other species may have resulted in a lack of adaptability. As a result, MDFWP managers began a search for a different strain of fish adapted to alpine lake conditions and compatible with native fish stocks.

Brood-Stock Development and Performance

The search for a new brood stock ended at McBride Lake, a small (9.3 hectares), high-elevation (1,999 m) lake in the northeast corner of Yellowstone National park, not far from the Beartooth Plateau. McBride Lake is a productive lake supporting a naturally reproducing population of Yellowstone cutthroat trout. Specimens from the lake have been confirmed electrophoretically as genetically pure Yellowstone cutthroat trout, in spite of a 1936 plant of rainbow trout in the lake (Varley 1981).

In 1969, gametes from 15 spawning pairs of McBride Lake cutthroat trout were brought into the MDFWP hatchery system. Outbreaks of furunculosis disease delayed production of fish for stocking until 1975. Since then, the McBride Lake strain has been the major strain used for alpine lake management in the area of Montana east of the Continental Divide.

Several steps have been taken to insure the maintenance of genetic variability in the McBride Lake strain brood stock. Brood fish are tested annually for genetic variability and compared to the wild population. Mating and selection of fish for future brood stock are done randomly. Eggs from each spawning are included in the brood group in the proportion that spawning is expected to contribute to the total. Selection of fish for brood purposes is done by dipnetting approximately equal numbers of fish and allocating each net load to brood or nonbrood groups by means of a predetermined random pattern. This methodology avoids inadvertent selection as described by Kapuscinski and Jacobson (1987). Electrophoretic testing of progeny of the original 15 pairs showed no detectable loss of variability after three generations in the hatchery; however, wild gametes from McBride Lake are reintroduced into the brood stock every third generation to maintain genetic diversity.

McBride Lake strain Yellowstone cutthroat trout have performed extremely well in alpine lakes. Naturally reproducing populations have been established in lakes containing inlets or outlets suitable for spawning. One of these, Marker Lake, at 3,313 m is the highest elevation fishery in Montana. Often, a single introduction is sufficient to generate a naturally reproducing population. Rapid growth occurs in productive lakes, and maximum lengths of 60 cm and weights of 3 kg have been documented. In less productive lakes, fish from the McBride Lake strain normally reach lengths of 25–30 cm in 3 or 4 years. Catchability is high, a common characteristic of cutthroat trout (Johnson and Bjornn 1978). Longevity is normally in the range of 6 to 7 years, but may reach 9 years.

McBride Lake strain cutthroat trout appear to be very adaptable in their food habits. Large *Daphnia* spp. and Diptera pupae provide the major portion of the diet in most alpine lakes. Terrestrial insects may be seasonally important. Hensler (1987) found fish of the McBride Lake strain in Axolotl Lake 2, Montana, preyed extensively on mottled sculpins *Cottus bairdi.*

Recent introductions of the McBride Lake strain into lowland lakes and reservoirs appear promising as an alternative management strategy. Several Montana reservoirs require annual stocking of large numbers of hatchery rainbow trout to maintain a fishery. In spite of the availability of good quality spawning habitat, rainbow trout stocks have rarely established naturally reproducing populations. If naturally reproducing populations of cutthroat trout can be established, diversification and improvement of reservoir fisheries may occur, with the potential for long-term reductions in fisheries management costs. Trojnar and Behnke (1974) demonstrated increased game fish biomass and angler opportunity can be produced when different species or strains can partition the available habitat. More time will be required to assess the ability of the McBride Lake strain to coexist with other fish species

Discussion

The success of McBride Lake strain Yellowstone cutthroat trout in the MDFWP alpine lakes management program is predictable given the fishes' adaptation to an alpine lake ecosystem. Adaptability of the strain to a variety of situations probably results from a relatively short evolutionary history (invasion of the upper Yellowstone drainage occurred within the last 8,000 years) in an environment favoring an ecological generalist rather than a specialist (Behnke 1979).

McBride Lake strain cutthroat trout have proven highly adaptable to a variety of single-species lacustrine ecosystems. However, they evolved in a simple ecosystem without competition from other species. Tests of their ability to successfully coexist with other species are still in progress.

Implications of the MDFWP McBride Lake strain program are clear for fisheries managers contemplating a move towards strain management. A program is likely to be successful if the strain selected is well adapted to its intended use. The MDFWP program is designed to take advantage of the adaptability of the McBride Lake strain in a variety of lake environments. A key element in the program is insuring retention of the genetic variability of a wild population and, consequently, the adaptability that made the strain well suited to its environment. In simple terms, a successful strain management program is one that takes advantage of natural selection that occurs in a wild population and attempts to avoid artificial selection for traits that favor hatchery performance.

References

Allendorf, F. W., and S. R. Phelps. 1980. Loss of genetic variation in a hatchery stock of cutthroat trout. Transactions of the American Fisheries Society 109:537–543.

Behnke, R. J. 1979. The native trouts of the genus *Salmo* of western North America. Report to U.S. Fish and Wildlife Service, Denver, Colorado.

Doyle, R. W. 1983. An approach to the quantitative analysis of domestication selection in aquaculture. Aquaculture 33:167–185.

Gresswell, R. E., and J. D. Varley. 1988. Effects of a century of human influence on the cutthroat trout of Yellowstone Lake. American Fisheries Society Symposium 4:45–52.

Hensler, M. E. 1987. A field evaluation of four strains of *Salmo* introduced into seven Montana waters. Master's thesis. Montana State University, Bozeman.

Johnson, T. H., and T. C. Bjornn. 1978. The St. Joe River and Kelly Creek cutthroat trout populations: an example of wild trout management in Idaho. Pages 39–47 *in* J. R. Moring, editor. Proceedings of the wild trout–catchable trout symposium. Oregon Department of Fish and Wildlife, Corvallis.

Kapuscinski, A. R., and L. D. Jacobson. 1987. Genetic guidelines for fisheries management. University of Minnesota, Minnesota Sea Grant Institute, St. Paul.

Trojnar, J. R., and R. J. Behnke. 1974. Management implications of ecological segregation between two introduced populations of cutthroat trout in a small Colorado lake. Transactions of the American Fisheries Society 103:423–430.

Varley, J. D. 1981. A history of fish stocking activities in Yellowstone National Park between 1881 and 1980. U.S. National Park Service, Information Paper 35, Yellowstone National Park, Wyoming.

American Fisheries Society Symposium 4:45–52, 1988

Effects of a Century of Human Influence on the Cutthroat Trout of Yellowstone Lake

ROBERT E. GRESSWELL

U.S. Fish and Wildlife Service, Post Office Box 184
Yellowstone National Park, Wyoming 82190, USA

JOHN D. VARLEY

U.S. National Park Service, Post Office Box 168
Yellowstone National Park, Wyoming 82190, USA

Abstract.—Over the past century, human activities have had an adverse influence on the Yellowstone cutthroat trout *Salmo clarki bouvieri* of Yellowstone Lake. Direct effects have included introductions of nonnative fishes, hatchery operations, and angler harvest. It appears that hatchery operations and angler harvest have been the most deleterious. Three of four nonnative fishes introduced to the system apparently filled empty niches in the lake ecosystem, and there is no evidence of direct competition between these introduced nongame fishes and native Yellowstone cutthroat trout. Potential competition from introduced nonnative brook trout *Salvelinus fontinalis* was thwarted by chemical removal of this species from Arnica Creek, a tributary to Yellowstone Lake. Although egg-taking and hatchery operations were terminated over 30 years ago, it is now apparent that egg removal, genetic mixing, and greatly reduced natural spawner escapement led to a gradual reduction in reproductive potential and undermined the complex mosaic of reproduction and recruitment. Angler harvest also had a negative impact as it became excessive. During the past decade, increasingly restrictive regulations have helped to restore population numbers and age structure. Current policies prohibit introductions of exotic species, emphasize natural wild stocks, and restrict angler harvests; these policies are products of a continued effort to manage Yellowstone Lake as an ecological unit.

Although Europeans first visited the area of Yellowstone National Park near the beginning of the 19th century, their presence did not begin to have a substantial effect on the area, and specifically on Yellowstone Lake, until the latter part of the century. Because of U.S. National Park Service policies, by-products of human activities that have degraded other fisheries, such as pollution and degenerative land use practices, have not been a factor in the Yellowstone Lake basin. Active suppression of lightning-caused fires from 1885 to 1972 (Sellers and Despain 1976), although ineffective until the late 1940s, reduced releases of nutrients to the lake and maintained a rather unproductive terrestrial vegetative mosaic throughout the watershed. These changes may have reduced lake productivity during this period; however, recent sediment analyses suggest that productivity changes within the past 200 years are related to climatic fluctuations (Val Klump, Center for Great Lakes Studies, personal communication).

As a native fish in an environment that has remained relatively pristine, the Yellowstone cutthroat trout *Salmo clarki bouvieri* of Yellowstone Lake provides an interesting study of direct human manipulation. Spawn-taking operations, commercial fishing, and sport fishery harvest have, singly and in combination, affected the Yellowstone cutthroat trout population. At various times, cutthroat trout stocks have been altered from their original state. To enable a better understanding of these effects, a brief introduction to the ecology of the lake is necessary before discussing these effects separately.

Yellowstone Lake and Fishery

Yellowstone Lake (Figure 1) lies at an elevation of 2,357 m. It is a large (35,391 hectare), oligotrophic lake with an average depth of 40 m (Benson 1961). Annual growing season is short; the lake usually freezes by late December and remains covered with about 1 m of ice until late May or early June. Summer surface temperatures rarely exceed 18°c.

Lodgepole pine *Pinus contorta* forests and subalpine meadows dominate the drainage basin, which has an estimated area of 261,590 hectares (Benson 1961). The bedrock geology is characterized by volcanic rock; andesite predominates in the eastern and southern portion of the drainage,

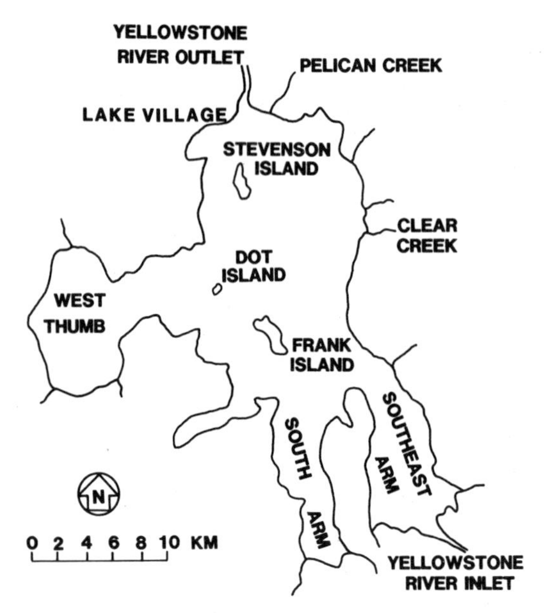

FIGURE 1.—Yellowstone Lake, Wyoming, and its major tributaries.

and rhyolite covers most of the west and north (USGS 1972). Chemically, it is a dilute sodium bicarbonate lake (total dissolved solids, 52–80 mg/L; U.S. Fish and Wildlife Service, Yellowstone National Park, unpublished data), typical of water originating from drainages of volcanic composition.

The siliceous nature of the watershed is apparent in the phytoplankton community. Diatoms are the major phytoplankton except for a brief pulse of the blue-green alga *Anabaena flos-aquae*, a nitrogen fixer, during periods of stratification (Benson 1961; Garrett and Knight 1973). Productivity is low; concentration of chlorophyll *a* in West Thumb, considered by Benson (1961) to be one of the most productive portions of the lake, is only 1–3 μg/L (Garrett and Kinght 1973). Production appears to be limited by nitrogen during the annual stratification and by water temperature and sunlight during the remainder of the year (Knight 1975).

The zooplankton is dominated by three species (*Diaptomus shoshone*, *Daphnia schødleri*, and *Conochilus unicornis*; Benson 1961). Despite relatively low zooplankton densities, these forms are important as food for juvenile Yellowstone cutthroat trout. Diets of mature Yellowstone cutthroat trout consist of zooplankton, two larger crustaceans (*Gammarus lacustris* and *Hyallela azteca*), and several families of Diptera, Ephemeroptera, and Trichoptera. Benson (1961) found that depth distribution, feeding, and movements of Yellowstone cutthroat trout were related to food abundance and spawning migrations.

Yellowstone Lake currently supports the largest inland population of cutthroat trout in the world. Although the only native fishes are Yellowstone cutthroat trout and longnose dace *Rhinichthys cataractae* (Simon 1962), two introduced species, redside shiner *Richardsonius balteatus* and longnose sucker *Catostomus catostomus*, are now abundant in the lake. The nonnative lake chub *Couesius plumbeus* is present in small numbers.

From May to July, mature Yellowstone cutthroat trout ascend 68 of the 124 known tributaries of Yellowstone Lake (Ball 1955; Cope 1956, 1957; Jones et al. 1986). Most are initial spawners (ages 4–6), although recidivism is common. Results of recent studies at Clear Creek suggest at least 26% of the spawners had spawned previously; repeat spawning was more common in alternate years (Jones et al. 1982, 1984). Larger Yellowstone cutthroat trout enter the stream first and appear to ascend farther up the tributaries than smaller individuals (Cope 1957; Ball and Cope 1961; Dean and Varley 1974). The amount of time that spawners remain in these streams varies from 1 to 3 weeks; usually males stay longer than females (Ball and Cope 1961). Ball and Cope (1961) reported a mean instream spawning mortality of 48.1% for five Yellowstone Lake tributaries between 1949 and 1953; however, recent estimates of instream spawning mortality at Clear Creek averaged only 12.9% (Jones et al. 1985).

Fry emerge about 30 d after egg deposition. Although some fry may remain in the natal stream for 1 or 2 years, most migrate to the lake within 2 months of emergence (Benson 1960). After leaving the tributaries, juveniles congregate along the shore before moving into deeper water. Because juveniles (through age 2) are rarely captured with gill nets set in shallow waters, it is assumed that pelagic waters of the lake serve as a nursery area.

Larger Yellowstone cutthroat trout (primarily mature) remain in the littoral zone throughout the year, and this distribution is especially evident during spawning migrations, when the fish appear to travel to and from spawning streams along the shoreline. This littoral orientation of mature Yellowstone cutthroat trout has played an important role in determining the mode of angler exploitation. Because most angling has been directed at littoral stocks, mature Yellowstone cutthroat trout are highly vulnerable to exploitation. Conversely, juvenile Yellowstone cutthroat trout (through age 2) have never entered the fishery in significant numbers because of their assumed pelagic distribution.

The Yellowstone Lake sport fishery has been studied intermittently throughout this century. Annual creel survey data have been gathered since 1950 (Moore et al. 1952; Jones et al. 1987). These long-term records of angler use, effort, and harvest provide an excellent data base for assessing the impact of the sport fishery on the Yellowstone cutthroat trout of Yellowstone Lake.

Nonnative Species Introductions

From the standpoint of preservation of native Yellowstone cutthroat trout, officially sanctioned attempts to introduce nonnative fish species into Yellowstone Lake were fortunate failures. Varley (1981) reported that mountain whitefish *Prosopium williamsoni* were stocked in the Yellowstone River below the lake in 1889 and 1890 (2,000 and 10,000 fish, respectively). Rainbow trout *Salmo gairdneri* were stocked twice in the lake (3,000 in 1902 and 3,800 in 1907) and once in the Yellowstone River (an unknown number of fingerlings in 1929). Two stockings of landlocked Atlantic salmon *Salmo salar* (7,000 in 1908 and 5,000 in 1909) were also unsuccessful.

Some fish introductions, officially sanctioned or not, were successful. Benson (1961) reported that records from the fish cultural station at Lake Village indicated that the longnose sucker was first introduced in 1923 or 1924. Redside shiners were first collected from the lake in 1957 (Cope 1958). Also in 1957, lake chub were reported in Indian Pond (within 300 m of Yellowstone Lake), but they were not collected from Yellowstone Lake until several years later (Cope 1958). Although it has long been assumed that these introductions were made by release of bait fishes by anglers, at least one source (Kelly 1928) stated that the longnose sucker may have been part of an authorized stocking of forage "minnows" in 1923.

Longnose suckers have been found in all portions of the lake, including the waters around the three major islands; however, redside shiners are restricted to the littoral zone (Biesinger 1961; Dean 1972). Lake chubs are most numerous in lagoons located in the northern part of the lake and West Thumb (Biesinger 1961). Native longnose dace seem to be confined to areas near the mouths of tributary streams.

Although populations of the longnose dace and lake chub are believed to be small and inconsequential, the longnose sucker and redside shiner have been the subject of studies concerned with their possible competition with Yellowstone cutthroat trout. An increase in numbers and distribution of longnose suckers between the early 1950s and the late 1960s was concomitant with a reduction in mean lengths and in angler landing rates of Yellowstone cutthroat trout.

Despite a certain amount of diet overlap (Benson 1961; Biesinger 1961), there is currently no evidence of direct competition between Yellowstone cutthroat trout and these introduced species (Jones et al. 1980; Swanson 1981). Recent improvements in the size structure of the Yellowstone cutthroat trout population have occurred since longnose suckers and redside shiners became established; this suggests that observed fluctuations in the status of the Yellowstone cutthroat trout have been influenced by factors other than the introduction of these nonnative fishes. Apparently, spatial separation within the lake and both temporal and spatial separation in spawning streams have led to a stable association of these fishes during the past decade. Marrin and Erman (1982) found that competition between trout (brown trout *Salmo trutta* and rainbow trout) and nongame species (tui chub *Gila bicolor* and Tahoe sucker *Catostomus tahoensis*) was "unlikely" in Stampede Reservoir, California. It seems that both the redside shiner and the longnose sucker have filled previously vacant niches within the Yellowstone Lake ecosystem.

An introduced population of brook trout *Salvelinus fontinalis* was discovered in a direct tributary to Yellowstone Lake (Arnica Creek) in May 1985 (Jones et al. 1986). These brook trout were considered a serious threat because introduced brook trout often replace native cutthroat trout (Griffith 1972; Hickman and Duff 1978), and within Yellowstone National Park, brook trout have been dominant in waters in which the two species coexist. Therefore, Arnica Creek and the Arnica Creek Lagoon were chemically treated to remove all fish in August 1985 (Jones et al. 1986), and treatment of Arnica Creek was repeated in August 1986 (Jones et al. 1987). Preliminary surveys in 1987 suggested that brook trout have been eradicated from the Arnica Creek drainage.

Hatchery Operations

The first egg collections at Yellowstone Lake occurred on streams in the West Thumb area in

1899, under supervision of personnel from Spearfish (South Dakota) National Fish Hatchery (Varley 1979). From this initial collection grew an immense operation that included a permanent hatchery at Lake Village and fish traps on 14 of the largest tributaries to the lake. From 1899 to 1957, more than 818 million eggs were gathered from Yellowstone cutthroat trout spawners entering these spawning streams (Varley 1979). A single-year high of 43.5 million eggs was taken in 1940. Although many of the resulting fry were returned to the lake, eggs and fry were shipped worldwide, and the Yellowstone Lake hatchery became what may have been the world's leading supplier of an inland wild trout species.

It is unfortunate that a program conceived with good intentions and carried out with a dedication for the resource became one of the most serious human threats to the cutthroat trout of Yellowstone Lake. The spawning runs in several of the larger tributary streams were blocked annually. Despite the return of fry to the lake, there was the potential of genetic mixing of up to 68 reproductive entities and the subsequent demise of smaller discrete spawning populations.

In some streams, reduced natural escapement of spawners (coupled with intense angler harvest) resulted in the virtual collapse of spawning migrations. At Clear Creek (Figure 1), for example, annual counts of spawners dropped from approximately 16,000 for the period between 1945 and 1948 to 3,353 spawners in 1954 (Benson and Bulkley 1963) (Figure 2). Numbers increased after spawn-taking activities ceased in 1953. Counts from 1957 to 1961 averaged 7,300 annually, and eight counts in the past decade have ranged from 32,000 to 70,000 cutthroat trout annually (mean, 47,800 cutthroat trout). The initial increase of spawners in the 1960s, which occurred during a period when the fishery was declining rapidly (Figures 3, 4), may be attributable to the termination of spawn taking.

Angler Use and Harvest

Except during wars and economic depressions, angling in Yellowstone Lake has increased steadily since the 1870s. Associated with the increase in effort were concomitant increases in annual harvest and declines in catch per unit effort. A daily creel limit of 20 fish was instituted in 1908 (Young 1908). Because of reports of a substantial decline in angler success, commercial fishing operations that supplied fresh cutthroat trout for hotel guests were discontinued in 1919 (Albright 1920). About 2 million Yellowstone cutthroat trout had been harvested by 1920, and our estimates suggest that over 7 million Yellowstone cutthroat trout had been harvested by 1954, a figure that may be unprecedented for an inland wild trout sport fishery.

In response to public and official concern during the 1950s, research biologists attempted to determine the number and sizes of Yellowstone cutthroat trout that could be harvested annually without deleterious effects to the population (Benson and Bulkley 1963). Management philosophy during that period was based on maximum sustained yield (MSY; Gresswell 1980). Estimates by Benson and Bulkley (1963) suggested that annual average MSY for Yellowstone Lake was

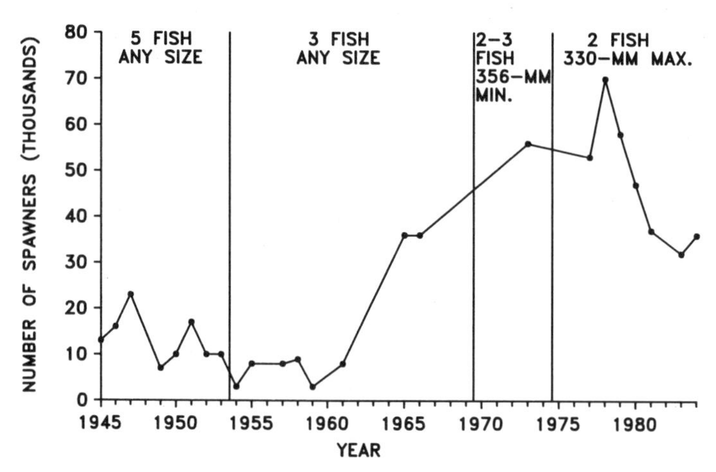

FIGURE 2.—Counts of cutthroat trout spawners at Clear Creek, Yellowstone Lake, under various angling restrictions, 1945–1984.

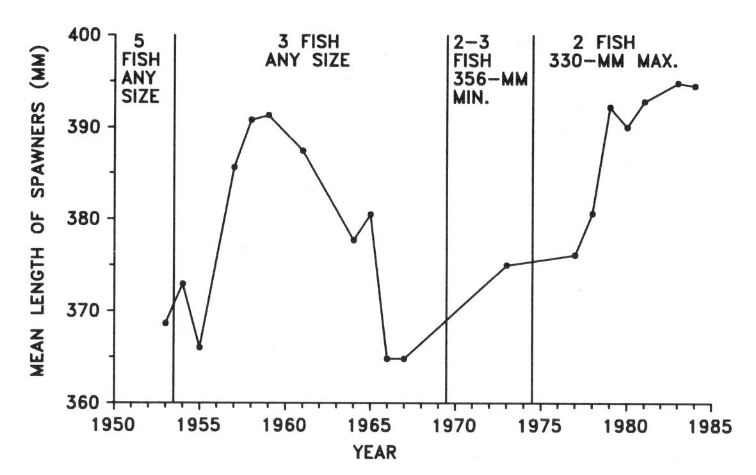

FIGURE 3.—Mean lengths of cutthroat trout spawners entering Clear Creek, Yellowstone Lake, under various angling restrictions, 1953–1984.

325,000 cutthroat trout (range, 290,000–340,000). It is now evident that these values were exceeded during the late 1950s and early 1960s (Figure 5). As a result of excessive harvest, the landing rate declined to a low of 0.45 cutthroat trout/h in 1969 (Figure 5). Mean lengths of cutthroat trout captured by anglers and trapped during spawning migrations decreased, and mean age for cutthroat trout spawners at Clear Creek dropped to a low of 3.9 years during the mid-1960s (Figures 3, 4, 6).

Beginning in 1969, increasingly restrictive angling regulations were imposed in effort to restore the ailing fishery (Gresswell, in press). Bait restrictions, reduction of the creel limit from three cutthroat trout to two, a 356-mm minimum-size limit, and closure of the Fishing Bridge Area to angling combined to foster stock recovery. Annual harvest dropped from 370,000 fish in 1958 to an average of 100,000 by 1974, and landing rate rose to about 1 cutthroat trout/h (Figure 5).

Despite signs of resurgence, it became evident that the 356-mm minimum-size limit was actually hastening the decrease of older and larger cutthroat trout in the population. This result was viewed with great apprehension because recruitment instability is often associated with single-age spawning stocks (Ricker 1954). Although estimates based on the Walford Line (Ricker 1975) indicated a historical maximum of 11 age-groups, Yellowstone cutthroat trout of age 6 and older made up only 6% of the catch by 1973. Experimental gill-netting data indicated that by 1974 the

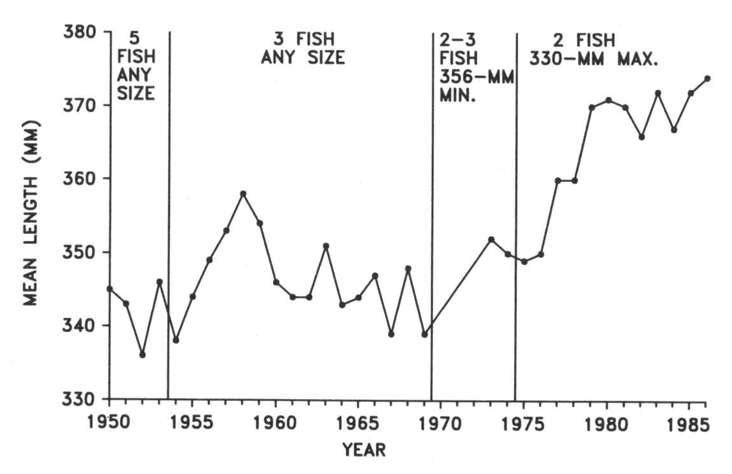

FIGURE 4.—Mean lengths of cutthroat trout captured by anglers under various angling restrictions, Yellowstone Lake, 1949–1986.

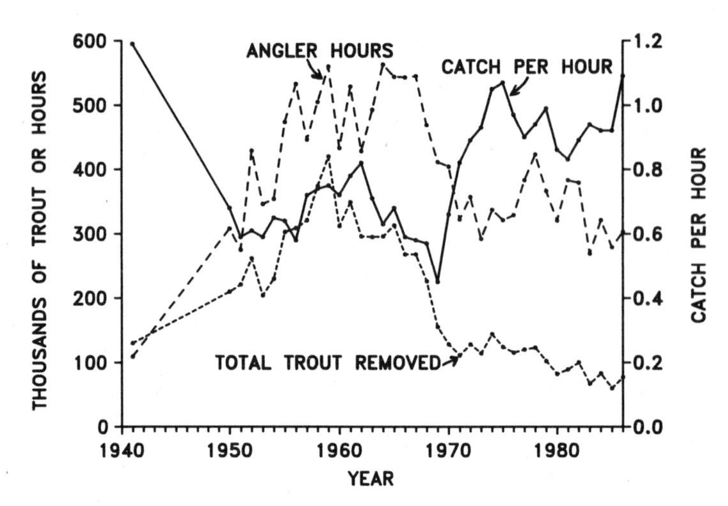

FIGURE 5.—Estimates of total angler effort (hours), thousands of cutthroat trout removed (including hooking mortality estimates after 1962), and numbers of cutthroat trout landed per hour for Yellowstone Lake, 1941 and 1950–1986.

average size of Yellowstone cutthroat trout prespawners had begun to decrease to levels observed before the minimum-size regulation was imposed in 1970 (Figure 7).

In order to prevent further deterioration of population structure and of the fishery in general, size-limit regulations were changed from a 356-mm minimum to a 330-mm maximum in 1975. Effects of the first 12 years of this regulation (1975–1986) are encouraging. Average age of Clear Creek spawners in 1983 was the highest yet reported (5.6 years), and age-9 Yellowstone cutthroat trout have begun to appear in the spawning migration annually (Figure 6). Average lengths of cutthroat trout landed in the fishery, spawners at Clear Creek, and prespawners in fall-set gill nets have continued to rise (Figures 3, 4, 7). Fecundity estimates suggest that current reproductive potential in Clear Creek (29.6 million eggs) is five times greater than it was during the 1950s (6.2 million eggs; Jones et al. 1985). These changes occurred while use (approximately 150,000 angler-days annually) and effort (over 350,000 angler-hours) remained high, and yet, the landing rate (averaging 0.94 cutthroat trout/h since 1975) is excellent. Currently five large cutthroat trout (356–457 mm) are captured and released for every cutthroat trout under 330 mm creeled.

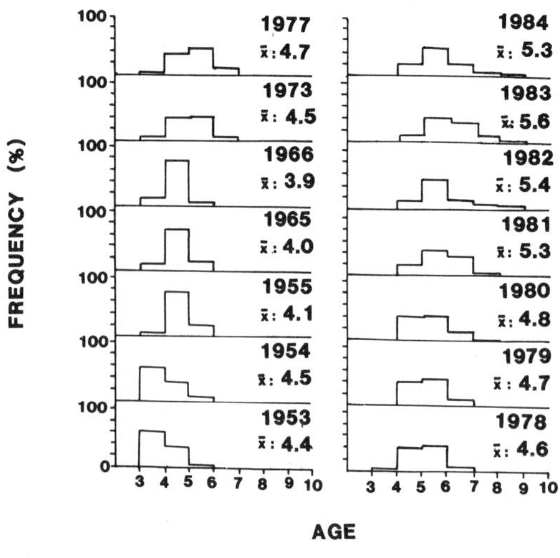

FIGURE 6.—Age-distribution histograms and mean ages of cutthroat trout spawners at Clear Creek, Yellowstone Lake, during various sampling years from 1953 to 1984.

Although the introduction of nonnative fishes remains a potential threat to the Yellowstone cutthroat trout population of Yellowstone Lake, the hatchery and egg-taking operation has been closed for over 30 years. Egg removal, genetic mixing, and greatly reduced natural spawner escapement all led to a gradual reduction in reproductive potential and undermined the complex mosaic of reproduction and recruitment. Combined with overharvest, this program led to the decline of the fishery during the early 1960s. Because the development of discrete, perhaps unique, genetic characteristics is measured in geologic time, the loss of genetic integrity may be virtually irretrievable.

Since the decline of the Yellowstone cutthroat trout fishery in Yellowstone Lake, increasingly

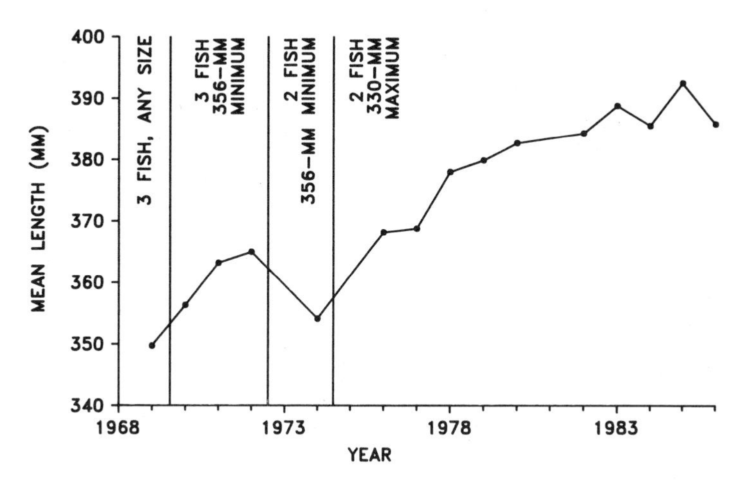

FIGURE 7.—Mean lengths of cutthroat trout prespawners (judged as ready to spawn the following spring) captured in September in experimental gill nets in Yellowstone Lake, during various angling restriction regimes, 1969–1986.

restrictive angling regulations have helped to restore population numbers and age-structure. Although the cutthroat trout population may yet be far from its historic potential, data indicate a state of equilibrium has been approached in recent years. Continued rise in angler use is cause for concern, however, because overharvest may occur even under regulations that allow only a minimal creel limit. The problem in the past, and one yet to be addressed, is the aspect of unlimited entry of anglers into the sport fishery.

Humans continue to have a substantial effect on the Yellowstone Lake cutthroat trout fishery. Data based on both historical records and experimental research are available for the continued evaluation of the Yellowstone Lake fishery. As we better understand past shortcomings, we can continually reassess management programs to ensure the proper protection for the resource while allowing high-quality angling experiences for park visitors. Current policies that allow natural fires, prohibit exotic species introductions, emphasize natural wild stocks, and restrict angler harvest are products of continued effort to manage the lake as an ecological unit.

Acknowledgments

This paper represents the work of a great number of U.S. Fish and Wildlife Service (USFWS), U.S. National Park Service (NPS), and Young Adult Conservation Corps employees, who in various ways aided in the collection of data. P. E. Bigelow, D. G. Carty, R. W. Eschmeyer, R. D. Jones, S. M. Rubrecht, and R. A. Valdez of the USFWS and D. G. Despain and M. M. Meagher of the NPS critically reviewed the manuscript. Additional comments by W. T. Helm, C. M. Kaya, R. L. Kendall, and R. F. Thurow were essential to the preparation of the final manuscript.

References

Albright, H. M. 1920. Superintendent's annual report for Yellowstone National Park 1919. U.S. National Park Service, Yellowstone National Park, Wyoming.

Ball, O. P. 1955. Some aspects of homing in cutthroat trout. Proceedings of the Utah Academy of Sciences, Arts, and Letters 32:75–80.

Ball, O. P., and O. B. Cope. 1961. Mortality studies on cutthroat trout in Yellowstone Lake. U.S. Fish and Wildlife Service Research Report 55.

Benson, N. G. 1960. Factors influencing production of immature cutthroat trout in Arnica Creek, Yellowstone Park. Transactions of the American Fisheries Society 89:168–175.

Benson, N. G. 1961. Limnology of Yellowstone Lake in relation to the cutthroat trout. U.S. Fish and Wildlife Service Research Report 56.

Benson, N. G., and R. V. Bulkley. 1963. Equilibrium yield management of cutthroat trout in Yellowstone Lake. U.S. Fish and Wildlife Service Research Report 62.

Biesinger, K. E. 1961. Studies on the relationship of the redside shiner (*Richardsonius balteatus*) and the longnose sucker (*Catostomus catostomus*) to the cutthroat trout (*Salmo clarki*) population in Yellowstone Lake. Master's thesis. Utah State University, Logan.

Cope, O. B. 1956. Some migration patterns in cutthroat trout. Proceedings of the Utah Academy of Science, Arts, and Letters 33:113–118.

Cope, O. B. 1957. Races of cutthroat trout in Yellowstone Lake. U.S. Fish and Wildlife Service Special Scientific Report—Fisheries 208:74–84.

Cope, O. B. 1958. New fish records from Yellowstone. Yellowstone Nature Notes 32(2):15–17. (U.S. National Park Service, Yellowstone National Park, Wyoming.)

Dean, J. L. 1972. Fishery management investigations, Yellowstone National Park. U.S. Fish and Wildlife Service, Annual Project Report, Yellowstone National Park, Wyoming.

Dean, J. L., and J. D. Varley. 1974. Yellowstone fishery investigations. U.S. Fish and Wildlife Service, Annual Project Report, Yellowstone National Park, Wyoming.

Garrett, P. A., and J. C. Knight. 1973. Limnology of the West Thumb of Yellowstone Lake, Yellowstone National Park. Montana State University, Final Report (Contract 2-101-0387), Bozeman.

Gresswell, R. E. 1980. Yellowstone Lake—a lesson in fishery management. Pages 143–147 *in* W. King, editor. Wild trout II. Trout Unlimited and Federation of Fly Fishermen, Vienna, Virginia.

Gresswell, R. E. In press. Response of Yellowstone Lake cutthroat trout to a 330-mm maximum size limit: 12 years of success. *In* R. A. Barnhart and T. D. Roelofs, editors. Catch-and-release fishing: a decade of experience. Humbolt State University, Arcata, California.

Griffith, J. S., Jr. 1972. Comparative behavior and habitat utilization of brook trout (*Salvelinus fontinalis*) and cutthroat trout (*Salmo clarki*) in small streams in northern Idaho. Journal of the Fisheries Research Board of Canada 29:265–273.

Hickman, T. J., and D. A. Duff. 1978. Current status of cutthroat trout subspecies in the western Bonneville basin. Great Basin Naturalist 38:193–202.

Jones, R. D., P. E. Bigelow, R. E. Gresswell, and R. A. Valdez. 1982. Fishery and aquatic management program, Yellowstone National Park. U.S. Fish and Wildlife Service, Technical Report for 1981, Yellowstone National Park, Wyoming.

Jones, R. D., D. G. Carty, R. E. Gresswell, K. A. Gunther, L. D. Lentsch, and J. Mohrman. 1985. Fishery and aquatic management program in Yellowstone National Park. U.S. Fish and Wildlife Service, Technical Report for 1984, Yellowstone National Park, Wyoming.

Jones, R. D., D. G. Carty, R. E. Gresswell, C. J. Hudson, L. D. Lentsch, and D. L. Mahony. 1986. Fishery and aquatic management program in Yellowstone National Park. U.S. Fish and Wildlife Service, Technical Report for 1985, Yellowstone National Park, Wyoming.

Jones, R. D., D. G. Carty, R. E. Gresswell, C. J. Hudson, and D. L. Mahony. 1987. Fishery and aquatic management program in Yellowstone National Park. U.S. Fish and Wildlife Service, Technical Report for 1986, Yellowstone National Park, Wyoming.

Jones, R. D., R. E. Gresswell, K. A. Gunther, and L. D. Lentsch. 1984. Fishery and aquatic management program in Yellowstone National Park. U.S. Fish and Wildlife Service, Technical Report for 1983, Yellowstone National Park, Wyoming.

Jones, R. D., J. D. Varley, D. E. Jennings, S. M. Rubrecht, and R. E. Gresswell. 1980. Fishery and aquatic management program in Yellowstone National Park. U.S. Fish and Wildlife Service, Technical Report for 1979, Yellowstone National Park, Wyoming.

Kelly, H. M. 1928. A sucker taken in Yellowstone Lake. Yellowstone Nature Notes 5(8):8. (U.S. National Park Service, Yellowstone National Park, Wyoming.)

Knight, J. C. 1975. The limnology of the West Thumb of Yellowstone Lake, Yellowstone National Park, Wyoming. Master's thesis. Montana State University, Bozeman.

Marrin, D. L., and D. C. Erman. 1982. Evidence against competition between trout and nongame fishes in Stampede Reservoir, California. North American Journal of Fisheries Management 2:262–269.

Moore, H. L., O. B. Cope, and R. E. Beckwith. 1952. Yellowstone Lake creel censuses, 1950–1951. U.S. Fish and Wildlife Service Special Scientific Report—Fisheries 81.

Ricker, W. E. 1954. Stock and recruitment. Journal of the Fisheries Research Board of Canada 11:559–622.

Ricker, W. E. 1975. Computation and interpretation of biological statistics. Fisheries Research Board of Canada Bulletin 191.

Sellers, R. E., and D. G. Despain. 1976. Fire management in Yellowstone National Park. Proceedings of the Annual Tall Timbers Fire Ecology Conference and Fire and Land Management Symposium 14:99–113. (Tall Timbers Research Station, Tallahassee, Florida.)

Simon, J. R. 1962. Yellowstone fishes, 3rd edition. Yellowstone Library and Museum Association, Yellowstone Interpretive Series 3, Yellowstone National Park, Wyoming.

Swanson, R. 1981. Some aspects of the biology of the longnose sucker in Yellowstone Lake, Yellowstone National Park, Wyoming. Master's thesis. University of Wyoming, Laramie.

USGS (U.S. Geological Survey). 1972. Geologic map of Yellowstone National Park. USGS, Miscellaneous Geological Investigations, Map I-711, Washington, D.C.

Varley, J. D. 1979. Record of egg shipments from Yellowstone fishes, 1914–1955. U.S. National Park Service, Informational Paper 34, Yellowstone National Park, Wyoming.

Varley, J. D. 1981. A history of fish stocking activities in Yellowstone National Park between 1881 and 1980. U.S. National Park Service, Informational Paper 35, Yellowstone National Park, Wyoming.

Young, S. B. M. 1908. Report of the superintendent of the Yellowstone National Park to the Secretary of the Interior. U.S. Government Printing Office, Washington, D.C.

American Fisheries Society Symposium 4:53–60, 1988

Westslope Cutthroat Trout in Montana: Life History, Status, and Management

GEORGE A. LIKNES

Montana Cooperative Fishery Research Unit, Biology Department, Montana State University Bozeman, Montana 59717, USA

PATRICK J. GRAHAM

Montana Department of Fish, Wildlife and Parks, 1420 East Sixth Avenue Helena, Montana 59620, USA

Abstract.—The historic range of westslope cutthroat trout *Salmo clarki lewisi* included western Montana, central and northern Idaho, a small portion of Wyoming, and portions of three Canadian provinces. The distribution and abundance of westslope cutthroat trout have drastically declined from its historic range during the last 100 years. Although previous studies in Montana identified strongholds of populations, the status of westslope cutthroat trout over its complete range is uncertain. Three life history forms are found within this range: (1) lacustrine–adfluvial stocks, (2) fluvial–adfluvial populations, and (3) fluvial fish. Migratory adults travel long distances during periods of high streamflow and spawn when water temperatures are near 10°C. Most migratory adults quickly leave the spawning grounds following the spawning act. Sexual maturity is attained at age 3 or older. In some drainages, repeat spawning occurs predominantly in alternate years. Westslope cutthroat trout are opportunistic in their food habits, but are not highly piscivorous; they tend to specialize as invertebrate feeders. The decline of westslope cutthroat trout has been attributed to overexploitation, genetic introgression, competition from or replacement by nonnative species, and habitat degradation. Westslope cutthroat trout occupy only 27.4% of their original range in Montana, and genetically pure populations are present in only about 2.5% of their historic range. As a whole, westslope cutthroat trout populations in Idaho and British Columbia, Canada, have fared better than in Montana. Management programs designed to protect the westslope cutthroat trout have helped to maintain or increase existing populations, and even to create new populations.

The decline in abundance and distribution of interior cutthroat trout *Salmo clarki* have been so dramatic that many subspecies are on the brink of extinction as genetically pure populations. It has been estimated that at least 99% of the original populations of interior cutthroat trout have been lost in the last 100 years (Behnke 1972). One of these subspecies, the westslope cutthroat trout *Salmo clarki lewisi* was once the dominant trout over a historic range that encompassed western Montana, central and northern Idaho, and a small portion of northwestern Wyoming in the USA and southwestern Saskatchewan, southern Alberta, and southeastern British Columbia in Canada (Figure 1); Behnke 1979). The Yellowstone cutthroat trout *S. c. bouvieri,* a subspecies found in adjacent drainages (Figure 1), has often been confused with the westslope cutthroat trout. Today the dominance of the westslope cutthroat trout has waned, and its present status is uncertain. Protection of the remaining populations is essential because, in addition to being uniquely adapted to a particular water, each population represents an exclusive source of genetic material for increasing diversity of brood stocks used for propagation and reintroductions.

The name "westslope" suggests a presence west of the Continental Divide, but a considerable portion of this subspecies' original range lies east of the Continental Divide. The current and historic ranges of the westslope cutthroat trout encompass the upper Missouri River drainage upstream from Fort Benton, Montana, as well as the headwaters of the Marias, Judith, Musselshell, and Milk rivers, which are tributaries that enter the Missouri River below Fort Benton (Behnke 1979). This subspecies is also found in the South Saskatchewan River system south of the Bow River in the Hudson Bay drainage (Behnke 1979). Cutthroat trout in the upper Missouri River drainage are often referred to as upper Missouri cutthroat trout, but they are essentially identical to the westslope cutthroat trout of the upper Columbia River and to those in the South Saskatchewan River drainage (Roscoe 1974; Phelps and Allendorf 1982). West of

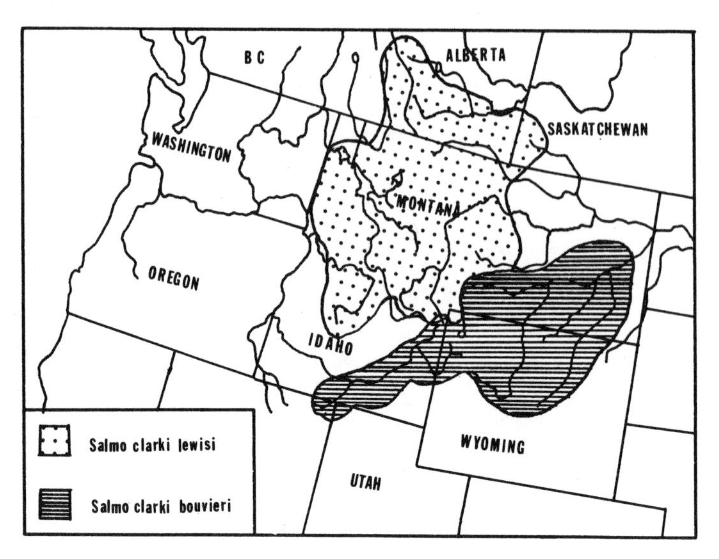

FIGURE 1.—The original ranges of the westslope cutthroat trout and the Yellowstone cutthroat trout in North America (Behnke 1979).

the Continental Divide, their range includes the Clark Fork River drainage upstream from the falls of the Pend Oreille River, the Kootenai River drainage from the headwaters to below the confluences of the Moyie and Elk rivers, the Spokane River basin above Spokane Falls, which includes the Coeur d'Alene and St. Joe river drainages and the Salmon and Clearwater river drainages (Behnke 1979).

The westslope cutthroat trout probably represents the first divergence of an interior cutthroat trout from the coastal cutthroat trout *S. c. clarki* ancestor (Behnke 1979). It evolved in sympatry with several other salmonids, including bull trout *Salvelinus confluentus* and mountain whitefish *Prosopium williamsoni,* throughout its range west of the Continental Divide. Below the barrier falls that isolated the upper Columbia basin, westslope cutthroat trout are found in waters containing Kamloops strain rainbow trout and steelhead *Salmo gairdneri* in some Idaho and British Columbia streams and chinook salmon *Oncorhynchus tshawytscha* in Idaho.

The presence of westslope cutthroat trout above barrier falls is indicative of a primitive dispersal pattern (Behnke 1979) and suggests that rainbow trout and chinook salmon expanded into the upper Columbia basin more recently than westslope cutthroat trout. The two native salmonids that coexisted with westslope cutthroat trout east of the continental divide were mountain whitefish and Arctic grayling *Thymallus arcticus.* Westslope cutthroat trout are thought to have crossed the Continental Divide from the west and expanded their range into the South Saskatchewan River and upper Missouri River systems about 7,000–10,000 years ago; the site of crossover from the Columbia River basin was probably in the vicinity of Glacier National Park, where the headwaters of all three drainages are adjacent (Roscoe 1987). It may have been relatively easy to move from one drainage to another during the Pleistocene when glacial lakes spanned drainage divides (Roscoe 1987).

The drastic decline of westslope cutthroat trout in portions of its historic range and its importance as a sport fish have resulted in efforts to prevent further loss of populations. Perturbations affecting some of these westslope populations include overexploitation, introgression, competition with introduced salmonids, and habitat degradation or loss. Each westslope cutthroat trout population is important because little genetic variation is present within a population but a large amount of variation occurs between populations (Leary et al. 1984). In the following discussion, we will examine (1) the life history patterns and habitat needs of various westslope cutthroat trout populations, (2) the present status of westslope cutthroat trout, (3) some of the ecological problems facing them, and (4) management programs designed to protect their remaining strongholds.

Life History

Three life history forms of westslope cutthroat trout are known to occur: lacustrine–adfluvial stocks, which migrate between lakes and streams; fluvial–adfluvial populations, which move between the main rivers and tributaries; and fluvial fish, which spend their entire lives in small headwater tributaries. Both lacustrine–adfluvial and fluvial–adfluvial fish will live from 2–3 years in tributary streams but they may spend as little as 1 or as long as 4 years in natal streams prior to migration. Behavioral rather than morphological differences may separate lacustrine–adfluvial and fluvial–adfluvial fish (Averett and MacPhee 1971); however, the distinctive genetic and behavioral features of these forms have not been fully explored.

Westslope cutthroat trout migrations of considerable magnitude occur within some systems. These include spawning and smolt migrations, downstream movements from tributary streams to overwinter, or simply movements to other portions of lakes or rivers, which may be related to food availability. Upstream spawning movements of 212 km in 87 d (2.44 km/d) have been reported (Shepard et al. 1984).

Juveniles emigrate downstream primarily at age 2 and 3 (Shepard et al. 1984). However, the number of age-1 outmigrants may also be substantial, and occasionally, individuals may be as old as age 4 before emigrating. Most juveniles move downstream during spring and early summer. Some juvenile westslope cutthroat trout may move out of natal streams, overwinter in a river, and then migrate to a lake (Shepard et al. 1984). Emigration has also been documented in the fall (Bjornn and Mallet 1964; May and Huston 1974, 1975) and may be a result of inadequate overwintering habitat. In such cases, early summer is a time of upstream migration for juveniles and adults that will reside in tributaries during the summer.

Fluvial westslope cutthroat trout and those juveniles that do not move out of small tributary streams may enter crevices in the substrate when water temperatures drop to 4–5°C. Summer is a time of little movement for westslope cutthroat trout; fish establish summer feeding stations which tend to define the primary behavioral pattern for that time period.

Timing of spawning activity is dependent on water temperature; adfluvial adults move into tributaries during high streamflows and spawn between March and July (Roscoe 1974), when water temperatures are near 10°C (Scott and Crossman 1973). Fluvial westslope cutthroat trout exhibit similar behavior. Most adfluvial adults spend little time in the tributaries, moving downstream shortly after spawning is completed. Spawning in some drainages occurs in the smaller tributaries, which may prevent mortality associated with streambed scouring that occurs in larger streams during high water conditions (Johnson 1963).

Westslope cutthroat trout in the Flathead River basin attain sexual maturity at age 4 and older, and 5- or 6-year-old fish from Coeur d'Alene Lake spawn in Wolf Lodge Creek (Lukens 1978); elsewhere, they may mature as early as age 3 (Brown 1971). Size of westslope cutthroat trout at maturity varies widely, but adult lacustrine–adfluvial fish tend to be the largest, usually greater than 350 mm total length (TL) (Shepard et al. 1984). Fluvial–adfluvial forms vary from 250 to 350 mm TL, and fluvial fish are generally less than 250 mm with some as small as 150 mm TL (Shepard et al. 1984). Spawning populations of westslope cutthroat trout tend to have a high ratio of females to males; the sex ratio from three Montana waters and one Idaho stream was 3.4:1 (Huston 1972; Lukens 1978; Huston et al. 1984; Shepard et al. 1984). Fecundity of westslope cutthroat trout appears to be slightly higher than for other subspecies and varies from 1,000 to 1,500 eggs for females with a mean length and weight of 355 mm and 0.5 kg, respectively (Roscoe 1974).

The proportion of repeat spawners tends to vary greatly, accounting for 0.7% of the fish in the 1975 run into Young Creek, Montana (May and Huston 1975); in Hungry Horse Creek, 24 and 19% of the spawners in 1970 and 1971, respectively, had spawned previously (Huston 1972, 1973). Low numbers of repeat spawners are associated with high spawning mortality (Scott and Crossman 1973). In some drainages, repeat spawning predominantly occurs in alternate years.

Westslope cutthroat trout in the Flathead River drainage spawn in areas where gravel varies from 2 to 50 mm in diameter, mean depths range from 17 to 20 cm, and mean velocities range between 0.30 and 0.37 m/s (Shepard et al. 1984). Fluvial westslope cutthroat trout built smaller redds than adfluvial fish. Redds varied from 0.6 to 1.0 m in mean length and from 0.32 to 0.45 m in mean width (Shepard et al. 1984). Eggs require about 310 temperature units to hatch (temperature units equal the sum of the mean daily temperatures above 0°C). Yolk-sac larva remain in the gravel until the yolk sac is absorbed (Shepard et al. 1984).

Young westslope cutthroat trout tended to be evenly distributed along stream margins in low-velocity areas such as pools and runs, but larger, older fish displayed a preference for pools (Shepard et al. 1984). Dominance hierarchies were usually maintained in pools where westslope cutthroat trout aligned vertically in the water column (Shepard et al. 1984). Maintenance of position in the water column requires much more energy than bottom orientation by species such as the bull trout (Shepard et al. 1984); however, the midwater strategy may allow the westslope cutthroat trout to obtain more food. This advantage may more than compensate for the greater expenditure of energy.

Westslope cutthroat trout fry emerge from the gravel at approximately 20 mm TL and first form scales at lengths between 38 and 44 mm (Shepard et al. 1984). In some areas, fry do not grow sufficiently to form annuli or scales the first year. From 61–69% of the westslope cutthroat trout sampled in the North and Middle Forks Flathead rivers failed to form annuli in the first year (Shepard et al. 1984). In the Coeur d'Alene River drainage, Lukens (1978) found that all fish examined had formed the first annulus.

Typical westslope cutthroat trout streams are cold, nutrient-poor waters where conditions for growth tend to be less than optimal. For example, the average lengths of westslope cutthroat trout from the upper Flathead River basin at each age were: age 1, 55 mm; age 2, 100 mm; age 3, 146 mm; age 4, 194 mm; age 5, 251 mm; and age 6, 301 mm (Shepard et al. 1984). In other portions of the present range of the westslope cutthroat trout, annual growth is greater. In tributaries of the St. Joe River, Idaho, the total lengths at each annulus for migrating fish were: age 1, 71 mm; age 2, 135 mm; and age 3, 226 mm (Thurow and Bjornn 1978). This is substantially slower growth than for the Yellowstone cutthroat trout (Varley and Gresswell 1988, this volume). Growth rates increase after fish emigrate from their natal tributaries; fish that moved from headwater streams in the Flathead River basin grew an average of 89–119 mm in the year after emigration but grew only 40–60 mm during years spent in tributaries (Shepard et al. 1984).

Westslope cutthroat trout tend to be opportunistic in their food habits and are not highly piscivorous; instead they tend to specialize as invertebrate feeders (Roscoe 1974; Behnke 1979). This dependence on invertebrates for food is attributed to their sympatric evolution with two highly piscivorous species, the bull trout and northern squawfish *Ptychocheilus oregonensis*. Utilization of aquatic insects in their diet may have prevented direct competition (Roscoe 1974; Behnke 1979) or may have been the result of competition.

Dipterans and ephemeropterans are the most important dietary components for westslope cutthroat trout; trichopterans are also an important dietary constituent for larger fish (110 mm long or longer; Shepard et al. 1984). Winged insects are not important in the diets of smaller fish (less than 110 mm) but become prominent as the fish increase in size. The diversity of food items used increases as the fish become larger. The most important food item for westslope cutthroat trout in Flathead and Priest lakes, which have large populations of planktivorus kokanee *Oncorhynchus nerka*, is terrestrial insects. Elsewhere, studies have shown that westslope cutthroat trout feed primarily on zooplankton as well as on terrestrial and aquatic insects (Jeppson and Platts 1959; Carlander 1969).

Population Status

The first report of westslope cutthroat trout by Europeans came from the journal of Lewis and Clark on 13 June 1805, when six cutthroat trout were caught while camped near the Great Falls of the Missouri. More recent surveys (Hanzel 1959) in Montana documented westslope cutthroat trout as the only or the predominant game fish in 230 streams and 130 lakes, but this effort was concentrated in the portion of the range east of the Continental Divide.

The historic range of the westslope cutthroat trout in Montana was conservatively estimated at 25,547 stream kilometers (Liknes 1984). The majority of this range, 54.4%, was west of the Continental Divide, 44.7% was east of the Continental Divide, and 0.9% was in the South Saskatchewan River system. Westslope cutthroat trout are currently known to be present in 655 streams and 6,993 stream kilometers in Montana or 27.4% of the historic range (Liknes 1984). These data include waters containing introgressed westslope cutthroat trout populations.

In 1984, 25 streams, consisting of 290 stream kilometers were known to contain genetically pure westslope cutthroat trout populations. Since

1984, 47 additional genetically pure populations have been documented; 30 were upgraded, previously classified streams, and 17 were previously unclassified and included many small first-order streams. These additional streams boost the proportion of genetically pure westslope cutthroat trout waters to over 2.5% of their historic range and 9.0% of their current range in Montana. Streams known to contain and those that potentially could contain genetically pure populations represented slightly more than 2,000 stream kilometers or about 29% of the current and 8% of the historic range (Liknes 1984).

Pure westslope cutthroat trout populations are known (or thought) to be present in nine river drainages in Montana. The upper Flathead River basin is the largest stronghold; westslope cutthroat trout are still present in 85% of their historic range in the upper Flathead River basin. The South Fork Flathead River is the largest and most secure portion within this area.

In Montana, 265 lakes are believed to contain westslope cutthroat trout populations; 8.3% of these lakes are known to contain genetically pure populations. Nineteen of the 22 genetically pure populations are in Glacier National Park (Marnell 1988, this volume); other lakes are in the Flathead River drainage. Only four lakes or reservoirs east of the Continental Divide in Montana were found to contain westslope cutthroat trout populations.

A status evaluation similar to that done in Montana has not been conducted in Idaho or Canada, but a greater proportion of the populations in those areas appears to have remained intact. Westslope cutthroat trout have been lost completely from some streams in Idaho, but many populations are genetically pure. In Idaho, important populations are found in the Middle Fork Salmon, North Fork Clearwater, Lochsa, Selway, St. Joe, and Coeur d'Alene rivers. However, the subspecies is present only in small numbers in most other streams of the Salmon River drainage and the South Fork Clearwater River.

Factors Affecting Population Abundance

Of all the factors threatening westslope cutthroat trout populations, hybridization with rainbow trout, golden trout *Salmo aguabonita,* and Yellowstone cutthroat trout represents the biggest problem. Because the westslope cutthroat trout was isolated and evolved separately from rainbow trout, it lacks an innate isolating mechanism which would allow the two species to coexist without hybridization.

Although a westslope cutthroat trout population may have undergone introgression, phenotypically it can still appear pure. Leary et al. (1983) demonstrated that identification by means of morphological characteristics does not accurately reflect the genetic composition of individuals or of a population of westslope cutthroat trout. Other tests using Yellowstone cutthroat trout and Alvord cutthroat trout have met with the same results (Tol and French, 1988, this volume; C. Clancy, Montana Department of Fish, Wildlife and Parks, personal communication).

Even though hybridization has occurred in westslope cutthroat trout populations, any population that closely resembles and has the same characteristics as the pure westslope cutthroat trout is still of considerable importance for management purposes. The trophy fishery of Ashley Lake in the Flathead River drainage is sustained by a rainbow trout × cutthroat trout hybrid population. Genetically pure populations of westslope cutthroat trout are still needed to reestablish or increase genetic variation in broodstocks maintained for propagation and reintroductions. Without genetic input from wild stocks, the highly selected hatchery fish could become increasingly vulnerable to disease, competition, predation, and changes in the physical environment. Wild, genetically pure populations also maintain the unique genetic-behavioral adaptations found in each individual population which have evolved in response to different environmental factors.

Although introgression has been widespread over most of the westslope cutthroat trout's historic range, there are several areas where populations of rainbow trout and westslope cutthroat trout coexist. The lower portion of the Flathead River above Flathead Lake contains rainbow trout, and at times, the same area is important in the life history of migratory westslope cutthroat trout. Rainbow trout have not greatly expanded their range within the drainage, and they have not hybridized extensively with westslope cutthroat trout. Evidently, spawning by the two species is both spatially and temporally isolated, and the drainage may be only marginal habitat for rainbow trout. In the Kootenai River drainage, different flow and temperature patterns of the Tobacco River tributaries and a difference in the timing of westslope cutthroat trout and rainbow trout spawning runs from Lake Koocanusa (Huston et al. 1984) tend to limit hybridization. Hybridization between the two species is widespread throughout the rest of the Lake Koocanusa system (Huston et al. 1984). Westslope cutthroat trout and steelhead in the Clearwater and Salmon river drainages have evolved sympatrically without significant hybridization. Mechanisms that limit the potential for hybridization between these species include aggressive spawning behavior and spatial separation of spawning sites.

The protection of high-quality habitat is essential for the continued existence of westslope cutthroat trout populations in streams. Habitat degradation is probably the second greatest influence (after hybridization with other salmonids) on westslope cutthroat trout populations. Human activities that can have detrimental effects include overgrazing by livestock, poor timber harvesting practices, oil and gas exploration, mining (placer operations in particular), water diversions, subdivisions and development of riparian zones, and construction of dams. Platts (1974) reported that cutthroat trout were common only in undisturbed reaches of streams in the Salmon River drainage of Idaho. Behnke (1979) described how clearcutting along two streams in the Smith River drainage of Montana increased erosion, sediment loads, and water temperatures; the westslope cutthroat trout population was eliminated in the disturbed area, and brook trout *Salvelinus fontinalis* was the principle species. A small area in the headwaters of one stream was not logged, and an indigenous westslope cutthroat trout population still dominated that reach. While each individual habitat perturbation may affect only a small portion of the historic range, the total of these habitat alterations have contributed greatly to the range reduction of the subspecies.

Westslope cutthroat trout populations are highly vulnerable to angling. MacPhee (1966) found that cutthroat trout were caught twice as easily as brook trout, and Behnke (1979) estimated that fishing pressure of 124 h/hectare per year will result in overexploitation of stream-dwelling westslope cutthroat trout populations. In contrast, brown trout *Salmo trutta* populations were not overexploited even with 1,235–1,976 h of fishing pressure per hectare per year (Behnke 1979). Because of their vulnerability to angling, westslope cutthroat trout populations depressed by overexploitation respond rapidly to special regulations, such as smaller bag limits, size restrictions, or catch-and-release fishing (Johnson and Bjornn 1978; Thurow and Bjornn 1978; Peters 1983). Five years after catch-and-release regulations were implemented on Kelly Creek, 13 times more cutthroat trout were counted in snorkeling transects (Johnson and Bjornn 1978). If the limiting factor is habitat-related, however, special regulations may cause a decrease of mortality due to angler harvest, but natural mortality will increase (Johnson and Bjornn 1978). Also, if anglers do not comply with regulations, depressed populations may not recover.

Competition with introduced salmonids is often listed as a major reason for the decline of cutthroat trout populations; however, there is a lack of detailed accounts and descriptions of the mechanisms involved (Liknes 1984; Griffith 1988, this volume). Although introduced salmonids may have actively displaced westslope cutthroat trout from some waters, brook trout and perhaps other introduced salmonid competitors may have simply replaced westslope cutthroat trout populations that had been depressed by other factors, such as high fishing pressure or habitat degradation (Griffith 1988). However, westslope cutthroat trout populations in high gradient areas at the heads of many streams may be able to remain essentially unaffected by introduced salmonids (Griffith 1988).

Management

The decline of westslope cutthroat trout in a specific water is usually not due to a single factor but rather to a combination of factors such as introgression, habitat degradation, exploitation, and competition or replacement. Each of these factors differs in importance for specific populations, and land managers and fisheries professionals must identify which management techniques can be best utilized to increase population levels.

Management steps taken for many waters include implementation of special regulations which maintain a resilient fishery resource and provide a quality fishing experience where catch rates, angler success, and satisfaction are high, and the mean length of fish caught is considered large. Species-specific fishing regulations that conserve westslope cutthroat trout populations but encourage harvest of introduced salmonids may benefit populations in some waters. Stream closures protect spawning areas of fluvial, fluvial–adfluvial, or lacustrine–adfluvial forms (Liknes 1984). After the implementation of special regulations or closures, monitoring must continue to verify that the measures taken are effective.

Informed management decisions regarding land and fishery use can be made only if the status of a westslope population is known. This information can be provided by a systematic program of electrophoretic analysis to identify genetically pure and introgressed populations. Other protective measures for westslope cutthroat trout populations include chemical rehabilitation to eliminate nonnative fishes, installation of physical fish barriers to prevent the expansion of nonnative fishes into westslope cutthroat trout waters, and acquisition of property, conservation easements, or water rights.

Management should emphasize the protection of existing strongholds, and policy decisions affecting those areas should be conservative and minimize uses that may adversely affect westslope cutthroat trout populations. The largest and most secure westslope cutthroat trout refuges in Montana are within wilderness areas and within Glacier National Park, where habitat degradation associated with human activities has been minimized. Maintenance of the wilderness system at present or expanded levels would prove beneficial to the westslope cutthroat trout.

Native salmonids such as the westslope cutthroat trout are often labeled "stupid" because their aggressive feeding behavior increases their vulnerability to angling. Such labels are counterproductive and should be avoided. Instead, a positive attitude, emphasizing high catch rates (which provide angler satisfaction) and the unique adaptations of a westslope cutthroat trout population for a particular water, should be stressed.

The current status of the westslope cutthroat trout appears more secure than several of the other interior cutthroat subspecies. Although the westslope cutthroat trout has disappeared from much of its historic range, we can maintain existing populations if we utilize wise conservation measures and manage with foresight.

References

Averett, R. C., and C. MacPhee. 1971. Distribution and growth of indigenous fluvial and adfluvial cutthroat trout, *Salmo clarki,* St. Joe River, Idaho. Northwest Science 45:38–47.

Behnke, R. J. 1972. The rationale of preserving genetic diversity: examples of the utilization of intraspecific races of salmonid fishes in fisheries management. Proceedings of the Annual Conference Western Association of Fish and Wildlife Agencies 52:559–561.

Behnke, R. J. 1979. The native trouts of the genus *Salmo* of western North America. Report to U.S. Fish and Wildlife Service, Denver, Colorado.

Bjornn, T. C., and J. Mallet. 1964. Movements of planted and wild trout in an Idaho river system. Transactions of the American Fisheries Society 93: 70–76.

Brown, C. J. D. 1971. Fishes of Montana. Big Sky Books, Montana State University, Bozeman.

Carlander, K. D. 1969. Handbook of freshwater fishery biology, volume 1. Iowa State University Press, Ames.

Griffith, J. S. 1988. Review of competition between cutthroat trout and other salmonids. American Fisheries Society Symposium 4:134–140.

Hanzel, D. A. 1959. The distribution of the cutthroat trout *(Salmo clarki)* in Montana. Proceedings of the Montana Academy of Sciences 19:32–71.

Huston, J. E. 1972. Life history studies of westslope cutthroat trout and mountain whitefish. Montana Department of Fish and Game, Federal Aid in Fish Restoration, Project F-34-R-5, Job III-a, Job Progress Report, Helena.

Huston, J. E. 1973. Life history studies of westslope cutthroat trout. Montana Department of Fish and Game, Federal Aid in Fish Restoration, Project F-34-R-6, Job III-a, Job Progress Report, Helena.

Huston, J. E., P. Hamlin, and B. May. 1984. Lake Koocanusa fisheries investigations, final completion report. Montana Department of Fish, Wildlife and Parks, Helena.

Jeppson, P. W., and W. S. Platts. 1959. Ecology and control of the Columbia River squawfish in northern Idaho lakes. Transactions of the American Fisheries Society 88:197–203.

Johnson, H. E. 1963. Observations on the life history and movement of cutthroat trout *(Salmo clarki)* in the Flathead River drainage, Montana. Proceedings of the Montana Academy of Sciences 23:96–110.

Johnson, T. H., and T. C. Bjornn. 1978. The St. Joe River and Kelly Creek cutthroat trout populations: an example of wild trout management in Idaho. Pages 39–47 in J. R. Moring, editor. Proceedings of the wild trout–catchable trout symposium. Oregon Department of Fish and Wildlife, Corvallis.

Leary, R. F., F. W. Allendorf, and K. L. Knudsen. 1983. Electrophoretic examination of trout from Lake Koocanusa, Montana: inability of morphological criteria to identify hybrids. University of Montana, Population Genetics Laboratory Report 83/8, Missoula.

Leary, R. F., F. W. Allendorf, and K. L. Knudsen. 1984. Population genetic structure of westslope cutthroat trout: genetic variation within and between populations. University of Montana, Population Genetics Laboratory Report 84/3, Missoula.

Liknes, G. A. 1984. The present status and distribution of the westslope cutthroat trout *(Salmo clarki lewisi)* east and west of the continental divide in Montana. Report to Montana Department of Fish, Wildlife and Parks, Helena.

Lukens, J. R. 1978. Abundance, movements, and age structure of adfluvial westslope cutthroat trout in the Wolf Lodge Creek drainage, Idaho. Master's thesis. University of Idaho, Moscow.

MacPhee, C. 1966. Influence of differential angling mortality and stream gradient on fish abundance in a trout–sculpin biotope. Transactions American Fisheries Society 95:381–387.

Marnell, L. F. 1988. Status of the westslope cutthroat trout in Glacier National Park, Montana. American Fisheries Society Symposium 4:61–70.

May, B., and J. E. Huston. 1974. Kootenai River fisheries investigations, phase 2, part 1. Montana Department of Fish and Game, Contract DACW 67-73-C-0003, Job Progress Report, Helena.

May, B., and J. E. Huston. 1975. Habitat development of Young Creek, tributary to Lake Koocanusa. Montana Department of Fish and Game, Contract DACW 67-73-C-0002, Final Job Report, Helena.

Peters, D. J. 1983. Rock Creek management survey. Montana Department of Fish, Wildlife and Parks, Federal Aid in Fish Restoration, Project F-12-R-29, Job II-a, Job Progress Report, Helena.

Phelps, S. R., and F. W. Allendorf. 1982. Genetic comparison of upper Missouri cutthroat trout to other *Salmo clarki lewisi* populations. Proceedings of the Montana Academy of Sciences 41:14–22.

Platts, W. S. 1974. Geomorphic and aquatic conditions influencing salmonids and stream classifiction with application to ecosystem classification. U.S. Forest Service, Billings, Montana.

Roscoe, J. W. 1974. Systematics of the westslope cutthroat trout. Master's thesis. Colorado State University, Fort Collins.

Roscoe, J. W. 1987. How the cutthroats reached Montana. Montana Outdoors 18(3):27–30.

Scott W. B., and E. J. Crossman. 1973. Freshwater fishes of Canada. Fisheries Research Board of Canada Bulletin 184.

Shepard, B. B., K. L. Pratt, and P. J. Graham. 1984. Life histories of westslope cutthroat and bull trout in the upper Flathead River basin, Montana. Montana Department of Fish, Wildlife and Parks, Helena.

Thurow, R. F., and T. C. Bjornn. 1978. Response of cutthroat trout populations to the cessation of fishing in St. Joe River tributaries. University of Idaho, Forest, Wildlife and Range Experiment Station Bulletin 25, Moscow.

Tol, D., and J. French. 1988. Status of a hybridized population of Alvord cutthroat trout from Virgin Creek, Nevada. American Fisheries Society Symposium 4:116–120.

Varley, J. D., and R. E. Gresswell. 1988. Ecology, status, and management of the Yellowstone cutthroat trout. American Fisheries Society Symposium 4:13–24.

American Fisheries Society Symposium 4:61–70, 1988

Status of the Westslope Cutthroat Trout in Glacier National Park, Montana

LEO F. MARNELL

U.S. National Park Service, Glacier National Park, West Glacier, Montana 59936, USA

Abstract.—Indigenous populations of westslope cutthroat trout *Salmo clarki lewisi* have been adversely affected in Glacier National Park by a succession of human interferences spanning more than half a century. Introductions and invasions of nonnative salmonids have altered ecological conditions (e.g., through predation and competition) for native cutthroat trout and have threatened the integrity of indigenous gene pools. Lacustrine populations of westslope cutthroat trout are imperiled by nonnative species through 84% of their original range in Glacier National Park. Introduced populations of Yellowstone cutthroat trout *Salmo clarki bouvieri* and hybrids of rainbow trout *Salmo gairdneri* × cutthroat trout *S. clarki* occur in 12 lakes distributed throughout the three continental drainages that have their headwaters in Glacier National Park. In spite of these perturbations, genetically unaltered populations of westslope cutthroat trout persist in 15 park lakes in drainages of the North and Middle Fork Flathead rivers where the species was historically present. Also, westslope cutthroat trout appear to have been transplanted into several small lakes draining to Lake McDonald (Middle Fork Flathead River). Native westslope cutthroat trout have not been found in the Missouri River or South Saskatchewan River drainages within the park boundary.

The headwaters of three continental drainages arise within Glacier National Park. Lakes and streams west of the Continental Divide discharge to the Flathead River and enter the Pacific Ocean via the Clark Fork of the Columbia River. A second divide east of the Continental Divide separates the headwaters of the South Saskatchewan River, which drains to Hudson Bay, from the Missouri River, which discharges to the Gulf of Mexico via the Mississippi River (Figure 1). Among the 23 species of fish native to Glacier National Park, several are indigenous to some portion of all three basins. These include the westslope cutthroat trout *Salmo clarki lewisi* and the mountain whitefish *Prosopium williamsoni.*

The indigenous fishery of Glacier National Park has been substantially altered from its pristine state during the past half century by introductions of nonnative fish. The problem has been compounded by the recent invasion of other nonnative species via rivers entering the park. The westslope cutthroat trout has been severely affected by these incursions.

The impacts of nonnative fishes on indigenous populations have recently been reviewed by Taylor et al. (1984); Marnell (1986); and Moyle et al. (1986). Effects of fish introductions on the westslope cutthroat trout in Glacier National Park include genetic contamination of some native populations, and ecological disturbances affecting several life history stages.

Recent studies by Marnell et al. (1987) have addressed the genetic concern. Conclusions and opinions expressed here about the ecological status of cutthroat trout populations in Glacier National Park derive from observations made during the course of that investigation. Specific goals of this paper are to (1) document the historic range of the westslope cutthroat trout in Glacier National Park, (2) chronicle the history of introductions and invasions of nonnative fishes, and (3) discuss the ecological impacts of nonnative species on westslope cutthroat trout.

Lacustrine populations of westslope cutthroat trout are emphasized for two reasons. First, the ability of the U.S. National Park Service (NPS) to manage unique biotas is greatly enhanced whenever such populations are (or are largely) resident within the jurisdictional boundary of a park. Cutthroat trout residing in the interior lakes of Glacier National Park are migratory to some extent; however, most appear to complete their life cycle in their natal drainage, spending their adult life in nearby lakes (Marnell et al. 1987).

Secondly, Glacier National Park is one of the last remaining enclaves where predominately lacustrine populations of westslope cutthroat trout survive anywhere throughout the subspecies' former range. This attaches special significance because lacustrine strains of this cutthroat trout subspecies could represent a unique genome. The existence of potentially important but genetically "invisible" traits has

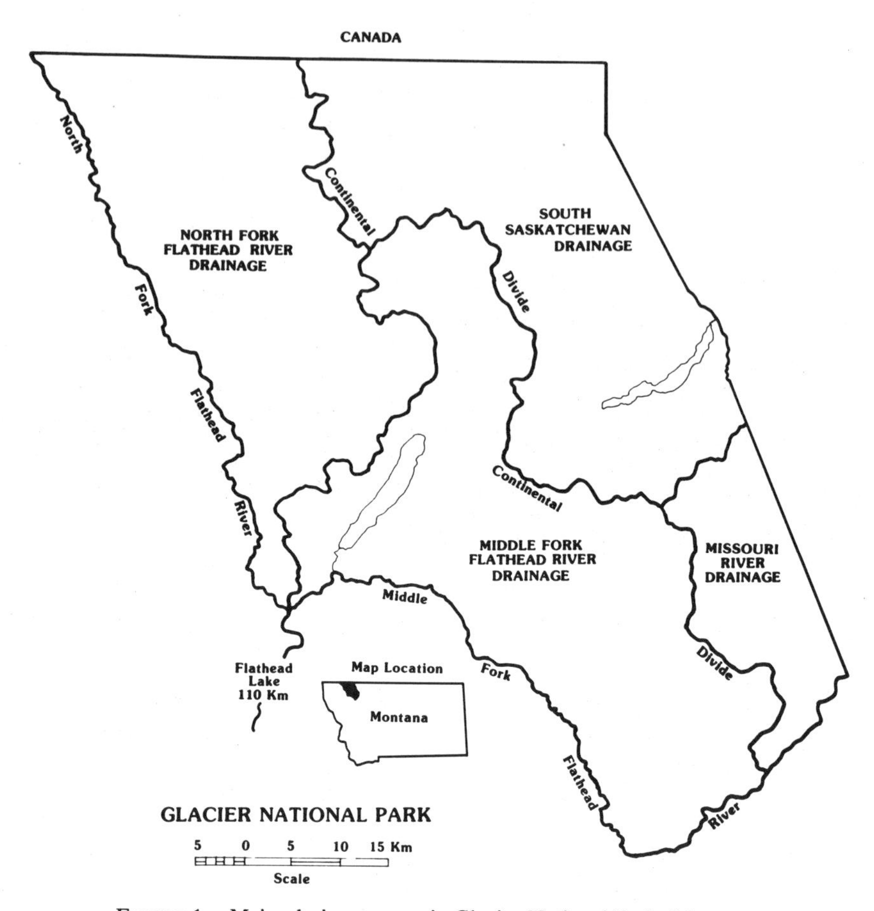

FIGURE 1.—Major drainage areas in Glacier National Park, Montana.

been postulated by a number of workers (Butler 1975; Crossman 1984; Marnell 1986).

Neither of these considerations diminishes the value or importance of fluvial westslope cutthroat trout in Glacier National Park. Several park drainages that are void of lakes provide important spawning and rearing habitats for fluvial westslope cutthroat trout and for recruitment to the adfluvial cutthroat trout fishery of Flathead Lake.

Historic Range

Westslope cutthroat trout attained dominance only in the low- to midelevation waters of the Flathead River in Glacier National Park. Natural barriers evidently prevented fish from reaching higher elevation lakes. Cerulean Lake (North Fork Flathead River; Figure 2) is the highest elevation water in the park known to contain indigenous cutthroat trout (elevation 1,420 m). Dispersal of this species from the upper Columbia River into the North Fork Flathead River and Middle Fork Flathead River (hereafter referred to as the North Fork and Middle Fork, respectively) drainages of Glacier National Park was reviewed by Marnell et al. (1987).

Westslope cutthroat trout probably gained access to the upper Missouri basin via headwaters crossover near Summit Lake, which spans the Continental Divide near the southern boundary of Glacier National Park (Figure 3). At one time this lake drained to both the east and the west. According to Schultz (1941), the westerly discharge to Bear Creek (headwaters of the Middle Fork drainage) was blocked by the Great Northern Railway during the early 1940s. Today, isolated populations of westslope cutthroat trout persist in some tributaries of the upper Missouri River outside the park as far downstream as the Musselshell River near Ft. Benton, Montana (Hanzel 1959).

There is no evidence that westslope cutthroat trout entered any of the interior lakes of Glacier National Park that drain to the upper Missouri

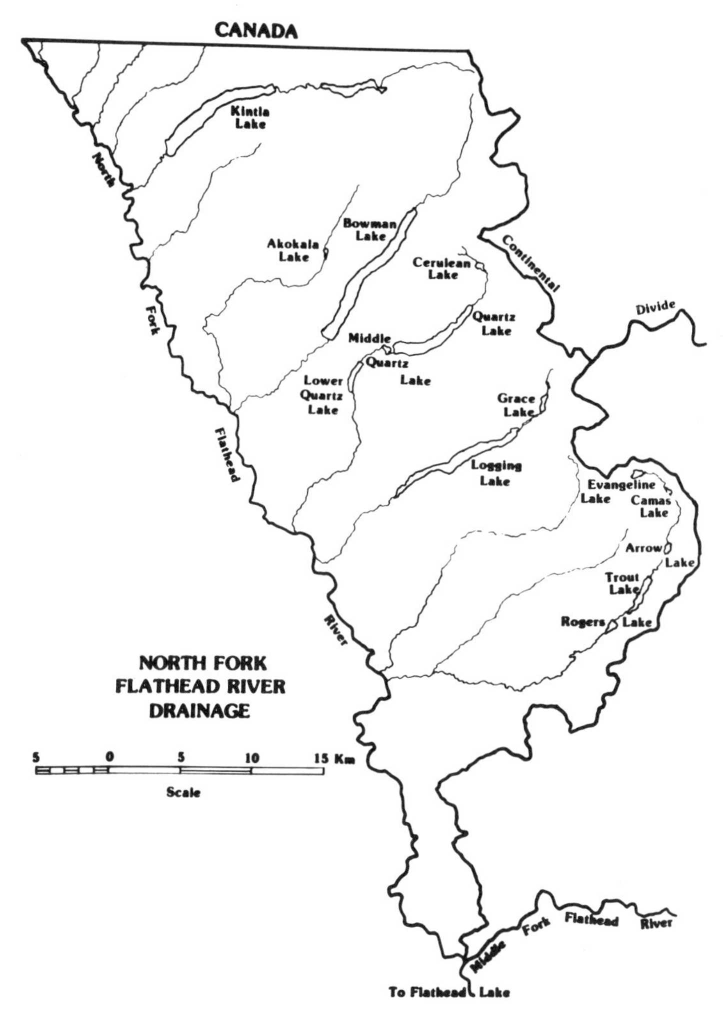

FIGURE 2.—Detail map of the North Fork Flathead River drainage in Glacier National Park, Montana.

River. Schultz (1941) believed, however, that the species may have been historically present in nearby Lower Two Medicine Lake. Construction of a dam outside the park in 1912 backed the reservoir inside the park boundary for about 2 km (Figure 3). It is unclear whether cutthroat trout were indigenous to Lower Two Medicine Lake because it has had heavy stocking with salmonids, including cutthroat trout. The same uncertainties apply to several small streams, including Cut Bank Creek, that exit the park in this region.

Dispersion of westslope cutthroat trout from the upper Missouri basin into the headwaters of the South Saskatchewan River may have been facilitated by ice dams; glacial ice-front lakes existed in the region during the retreat of the Wisconsin glaciers (Willock 1969). Schultz (1941) speculated, however, that more recent (i.e., post-glacial) faunal transfers may have resulted from drainage crossovers in the vicinity of Duck Lake (near St. Mary Village; Figure 4). Schultz believed some streams were simultaneously tributaries to the Milk River (upper Missouri River) and the St. Mary River (South Saskatchewan basin). It is also possible that a transcontinental crossing existed north of Canada's Waterton Lakes National Park, which adjoins Glacier National Park, directly connecting some reaches of the Columbia River headwaters with tributaries of the South Saskatchewan drainage.

Regardless of the route, westslope cutthroat trout entered the South Saskatchewan basin, including some midelevation waters of what is now Glacier National Park. While the species appar-

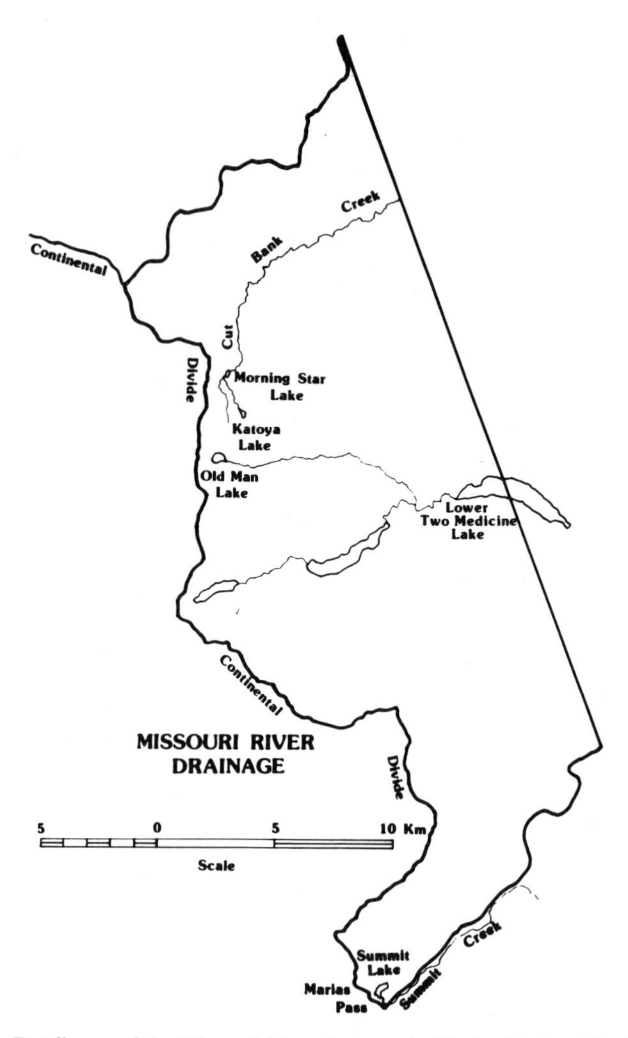

FIGURE 3.—Detail map of the Missouri River drainage in Glacier National Park, Montana.

ently did not disperse widely throughout this part of its range, small populations may have become established in the Waterton Valley, isolated reaches of the upper St. Mary River, and in the Belly River below Gros Ventre and Dawn Mist Falls (Figure 4). Schultz (1941) reported collections of cutthroat trout from the Belly River that he thought were native; however, none were found there during recent surveys (Marnell et al. 1987).

Introduction of Nonnative Fishes

Artificial enhancement of natural fisheries through stocking and the harvesting of fish by angling are fundamentally at odds with the mandate of the U.S. National Park Service to maintain natural ecosystems in an "unimpaired" (i.e., pristine) condition. This apparent contradiction owes its genesis, in part, to the fact that manipulation of aquatic ecosystems to accommodate sportfishing was already an established practice prior to the time that many of the national parks were established. Clearly, the doctrine of recreational fisheries management enjoyed wide support prior to creation of the National Park Service in 1916. The result of this de facto policy is that many valuable ecosystems have been irretrievably perturbed. Aquatic systems may represent the most compromised natural resource in the national parks today (Marnell 1985).

Early Fishery Manipulations

Alterations to the aquatic resource may have occurred prior to the establishment of Glacier National Park in May 1910. There are indications, for example, that early inhabitants of the region

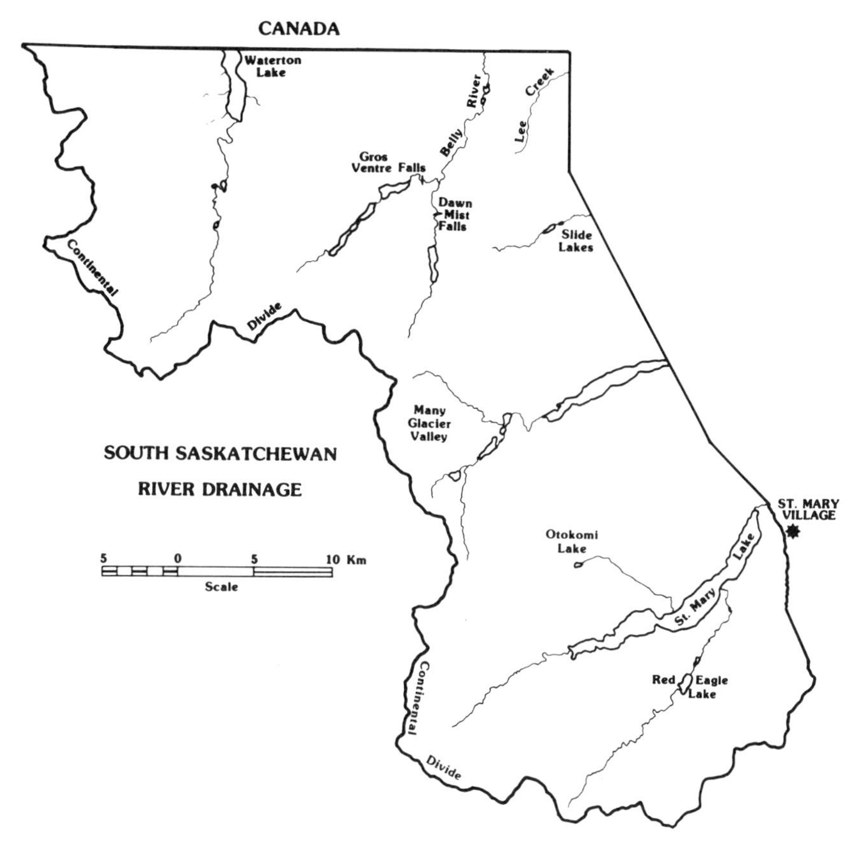

FIGURE 4.—Detail map of the South Saskatchewan River drainage in Glacier National Park, Montana.

may have transplanted wild cutthroat trout to Avalanche Lake and perhaps to Snyder, Fish, and the Howe lakes as well (Figure 5). These lakes drain to Lake McDonald via steep gorges containing numerous barriers; other salmonids usually resident with westslope cutthroat trout are absent. Although it is conceivable that only cutthroat trout were able to access these lakes, a more plausible interpretation is that wild fish were placed in them. Whatever their origin, cutthroat trout were evidently present prior to the existence of hatcheries anywhere in the region. This interpretation is supported by anecdotal accounts of trout being caught from these waters during the 1890s (NPS Archives, Glacier National Park). Despite later releases of Yellowstone cutthroat trout, only native westslope cutthroat trout are found in these waters today (Marnell et al. 1987).

Morton (1968) stated that several early residents obtained fish from the U.S. Fish Commission for stocking in Lake McDonald during the period 1910–1913. Also, the Great Northern Railway completed its transcontinental line in 1893 and actively promoted visitation to the region. It is likely that undocumented fish introductions were made by entrepreneurial interests in concert with the railroad's arrival.

Stocking of Hatchery-Reared Fish

National Park Service stocking records document fish stocking in Glacier National Park as early as 1912; however, most stocking took place during the period 1920–1955. Early introductions originated from a hatchery operated by the U.S. Fish Commission at East Glacier. Species stocked from this facility were primarily brook trout *Salvelinus fontinalis* and rainbow trout *Salmo gairdneri*.

Subsequent construction of hatcheries at Bozeman and Anaconda, Montana, led to the widespread introduction of Yellowstone cutthroat trout *Salmo clarki bouvieri* throughout the region. Most hatchery-reared cutthroat trout stocked in Glacier National Park prior to 1950 were the Yellowstone subspecies.

In 1947 a hatchery was constructed at Creston, Montana, exclusively for stocking park waters

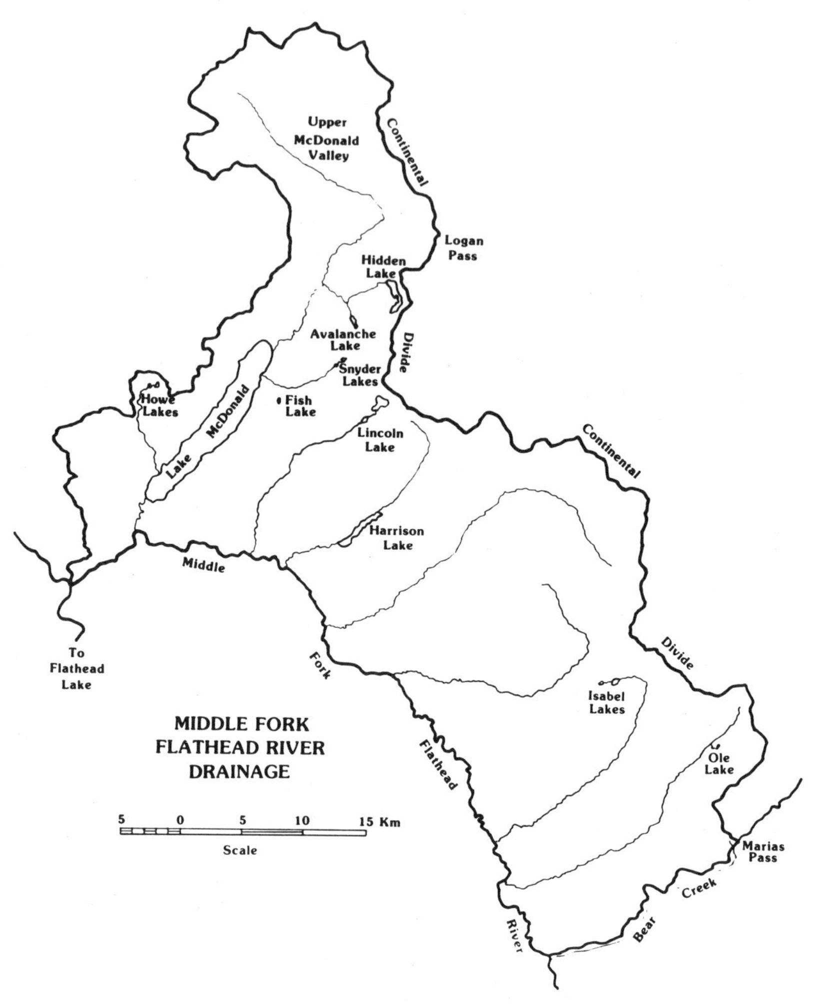

FIGURE 5.—Detail map of the Middle Fork Flathead River drainage in Glacier National Park, Montana.

with "native" cutthroat trout. Stocking from this facility continued into the 1960s, but diminished by the end of that decade in concert with a national trend toward reduced dependency on fish stocking in the national parks. Stocking was terminated in Glacier National Park in 1972.

This 70-year fish-stocking campaign resulted in the successful establishment of five nonnative salmonid species or subspecies in Glacier National Park. These included rainbow trout, brook trout, kokanee *Oncorhynchus nerka,* Yellowstone cutthroat trout, and Arctic grayling *Thymallus arcticus*. Nonnative populations, including various combinations of hybrids, were established in more than a dozen lakes where fish were previously absent (Marnell et al. 1987). All but four park lakes containing indigenous cutthroat trout (Cerulean, Lincoln, and Upper and Lower Isabel lakes) were at some time planted with rainbow trout or Yellowstone cutthroat trout. Both species are capable of hybridizing with the westslope cutthroat trout. Although Cerulean Lake received no plants, it drains directly to Quartz Lake, and the latter received a plant of Yellowstone cutthroat trout in 1948. Thus, only two indigenous trout populations (a single population inhabits Upper and Lower Isabel lakes) escaped direct exposure to potential introgression.

Invasion of Nonnative Species

During the first two decades of this century, lake trout *Salvelinus namaycush,* kokanee, and lake whitefish *Coregonus clupeaformis* were introduced by the state of Montana into several lakes throughout the Flathead River basin, including

Flathead and Whitefish lakes. An accurate chronology for the dispersal of these species is unavailable, but lake whitefish and kokanee had reportedly entered several of the large westslope lakes of Glacier National Park as early as the 1930s (Schultz 1941). There is no record that any of these species were stocked in the North or Middle Fork drainages inside Glacier National Park.

Schultz (1941) failed to collect lake trout from any park water west of the Continental Divide during his 1932–1934 surveys. This species evidently first appeared in McDonald Lake around 1960 (Harold Knapp, Missoula, Montana, personal communication). By the mid-1970s, lake trout accounted for nearly 50% of the fish caught from Lake McDonald (NPS creel census data, Glacier National Park). Lake trout have also invaded Bowman and Kintla Lakes in the North Fork drainage, and three were confirmed to have been caught during 1986 and 1987 from Logging Lake (identified by the author).

Current Status

Genetic contamination of native westslope cutthroat trout has been minimal in Glacier National Park despite the widespread stocking of closely related species and subspecies. This outcome is attributed to the superior ability of the native cutthroat trout to adapt to the harsh environs characteristic of the region. Most juvenile Yellowstone cutthroat trout released in park waters probably failed to survive to reproductive age. Contributing factors may have been downstream escapement of stocked cutthroat trout and the inability of introduced fish to coexist with native predators (e.g., bull trout *Salvelinus confluentus* and northern squawfish *Ptychocheilus oregonensis*). Failure of introduced Yellowstone cutthroat trout to adapt to the indigenous cestode parasite, *Ceratocephalus* sp. may also have been a factor (Marnell et al. 1987).

Ecological impacts associated with the introduction and invasion of nonnative fishes are less well defined. However, declines in westslope cutthroat trout populations in several park lakes may be related to predation or competition exerted by introduced species. The kokanee is known to be a particularly efficient open-water planktivore (Leathe and Graham 1982). Competition for zooplankton may be intense during winter when terrestrial insects are unavailable. Predation by invading lake trout may also be a significant cause of mortality among westslope cutthroat trout in these lakes.

North Fork Flathead River

The lakes and streams of the North Fork drainage represent the historical stronghold of the westslope cutthroat trout in Glacier National Park (Table 1). These populations are important because they reside mainly in lacustrine habitats, a potentially unique adaptation among surviving populations of this cutthroat trout subspecies. Among the 10 lakes that contained indigenous westslope cutthroat trout, only the Quartz Creek lakes and Akokala Lake appear to contain populations free of both genetic and ecological disturbances (Figure 2). The Akokala Lake specimens exhibit slight spotting irregularities but were nevertheless identified as genetically pure westslope cutthroat trout (Marnell et al. 1987).

National Park Service creel census data indicate that westslope cutthroat trout populations in Kintla and Bowman lakes have declined since the 1960s, evidently due to predation and competition from invading nonnative salmonids. Logging Lake will probably be similarly affected because kokanee and lake whitefish have been present there for more than a decade. The recent capture of several lake trout suggests that cutthroat trout in Logging Lake will probably experience increased predation.

Grace Lake, situated upstream from Logging Lake above a falls, contains a hybrid population of Yellowstone cutthroat trout × westslope cutthroat trout. This probably resulted from introductions of both cutthroat trout subspecies.

Arrow Lake (Camas Creek drainage) is the only water in Glacier National Park where Yellowstone cutthroat trout are known to have introgressed with an indigenous population of westslope cutthroat trout. Marnell et al. (1987) attributed this to continuous exposure from Yellowstone cutthroat trout entering Arrow Lake from Camas and Evangeline lakes (Figure 2).

Trout Lake and Rogers Lake (a shallow bog), situated further down the Camas Creek drainage, contain genetically pure populations of westslope cutthroat trout. Both, however, are subject to hybrid influence from upstream populations of Yellowstone cutthroat trout. While potential introgression with Yellowstone cutthroat trout remains a concern, other nonnative salmonids have evidently not entered the Camas Creek system.

Middle Fork Flathead River

The historic range of the westslope cutthroat trout in the Middle Fork drainage of Glacier

TABLE 1.—Status and distribution of cutthroat trout and their hybrids in the North and and Middle Fork Flathead river drainages in Glacier National Park, Montana.

Lake	Area (hectares)	Genetic identification[a]	Population origin	Nonnative species present[b]	Status of population
			North Fork Flathead River		
Kintla	688	(WCT)	Indigenous	LT, K, LWF	Moderately depressed
Akokala	9	WCT	Indigenous	None	Presently secure
Bowman	691	(WCT)	Indigenous	LT, K, LWF	Moderately depressed
Cerulean	20	WCT	Indigenous	None	Secure
Quartz	349	WCT	Indigenous	None	Secure
Middle Quartz	19	WCT	Indigenous	None	Secure
Lower Quartz	67	WCT	Indigenous	None	Presently secure
Grace	32	YCT × WCT	Introduced	None	Hybrids
Logging	444	WCT	Indigenous	LT, K, LWF	Decline anticipated
Evangeline	28	YCT	Introduced	None	Stable, nonnative
Camas	8	YCT	Introduced	None	Stable, nonnative
Arrow	23	WCT × YCT	Indigenous	YCT	Mildly hybridized
Trout	86	WCT	Indigenous	None	Presently stable
Rogers	34	(WCT)	Transient	None	Migratory
			Middle Fork Flathead River		
Upper Howe	3	WCT	Transplanted	RS	Secure
Lower Howe	12	WCT	Transpanted	RS	Secure
Hidden	110	YCT	Introduced	None	Stable, nonnative
Avalanche	23	WCT	Transplanted	None	Secure
Snyder	2	WCT	Transplanted	None	Secure
Fish	3	WCT × YCT	Introduced	None	Hybrids
McDonald	2,760	(WCT)	Indigenous	LT, K, LWF, RT, BT	Depressed
Lincoln	14	WCT	Indigenous	None	Secure
Harrison	101	WCT	Indigenous	K, BT	Presently stable
Upper Isabel	6	WCT	Indigenous	None	Secure
Lower Isabel	17	WCT	Indigenous	None	Secure
Ole	2	WCT	Indigenous	None	Stable

[a]Genetic identifications based on Marnell et al. (1987); WCT = westslope cutthroat trout; YCT = Yellowstone cutthroat trout. Genotypes appearing in parentheses are presumed correct, but not verified.
[b]LT = lake trout; K = kokanee; LWF = lake whitefish; RT = rainbow trout; BT = brook trout; RS = redside shiner *Richardsonius balteatus*.

National Park would be greatly diminished were it not for the large size (2,760 hectares) of McDonald Lake (Figure 5). However, NPS creel census data indicate that the cutthroat trout population in this water has declined in recent decades, apparently the result of ecological disturbances similar to those which have depressed westslope cutthroat trout populations in Kintla and Bowman lakes. The decline of this population represents a potential loss of up to 95% of the historic Middle Fork range of the westslope cutthroat trout in Glacier National Park.

Nine smaller lakes in the Middle Fork drainage also contain westslope cutthroat trout. Several associated with Lake McDonald are presumed to have been historically fishless and may harbor transplanted populations derived from wild stocks of westslope cutthroat trout (Table 1).

Lack of exposure to hybrid influence from hatchery fish attaches unique importance to the westslope cutthroat trout populations in Lincoln Lake and the Isabel lakes. Fish in the Isabel lakes are sympatric with a remnant population of dwarf bull trout and are isolated above a cascades. Thus, they are reasonably well protected from contact with nonnative species.

Harrison Lake contains westslope cutthroat trout in sympatry with introduced brook trout and kokanee. There is no indication that lake trout have entered this water. Ole Lake, a 2-hectare pond situated 26 km above the Middle Fork on Ole Creek, contains a small resident population of westslope cutthroat trout.

There are two populations of nonnative cutthroat trout in the Middle Fork drainage. Fish Lake (3 hectares), a disjunct pond near Lake McDonald, harbors a hybrid swarm of westslope cutthroat trout × Yellowstone cutthroat trout (Marnell et al. 1987). Hidden Lake (110 hectares) near Logan Pass contains Yellowstone cutthroat trout. It is the highest elevation water (1,943 m) in Glacier National Park inhabited by fish.

South Saskatchewan Drainage

The original distribution of westslope cutthroat trout in the South Saskatchewan River drainage of

TABLE 2.—Distribution of nonnative cutthroat trout and their hybrids in waters east of the Continental Divide, Glacier National Park, Montana, Fish were historically absent from these lakes.

Lake	Area (hectares)	Genetic identification[a]
South Saskatchewan River drainage		
Otokomi	9	YCT × RT
Upper Slide[b]	5	YCT × RT
Lower Slide[b]	15	YCT × RT
Red Eagle	55	RT × YCT
Upper Missouri River drainage		
Katoya	4	YCT
Old Man	17	YCT
Morning Star	4	YCT

[a]Genetic identifications based on Marnell et al. (1987); YCT = Yellowstone cutthroat trout; RT = rainbow trout.
[b]Upper and Lower Slide lakes are inhabited by a single population.

Glacier National Park appears to have been limited (Schultz 1941). Widespread stocking of rainbow trout and Yellowstone cutthroat trout probably resulted in hybridization or the loss of any indigenous populations that may have existed. Reports are occasionally received of cutthroat trout being taken from the Belly River (D. Shea, Glacier National Park, personal communication); these are probably westslope cutthroat trout × rainbow trout hybrids because introduced rainbow trout occur in the headwaters of the drainage.

Marnell et al. (1987) has identified populations of rainbow trout × Yellowstone cutthroat trout hybrids in Upper and Lower Slide lakes (single population) and in Otokomi Lake. These lakes collectively embody about 29 hectares (Figure 4). Red Eagle Lake (55 hectares) harbors a population of predominately rainbow trout introgressed with Yellowstone cutthroat trout (Table 2).

The absence of westslope cutthroat trout in Waterton Lake is puzzling. Waterton Lake (385 hectares) is clearly within the range of this species and contains several native species of fish typically associated with cutthroat trout. The absence of westslope cutthroat trout could be related to the presence of lake trout, considered by some workers (Schultz 1941; Morton 1968) to be indigenous to Waterton Lake.

Remnant populations of cutthroat trout may still exist in isolated portions of the South Saskatchewan River basin inside Glacier National Park. Although recent investigations have emphasized lacustrine populations of westslope cutthroat trout, the search for native trout in this drainage has now shifted to protected reaches of small streams.

Upper Missouri River Drainage

Introduced Yellowstone cutthroat trout reside in Katoya, Old Man, and Morning Star lakes, collectively encompassing 25 hectares (Figure 3; Table 2). The populations are stable and self-sustaining. They have probably been able to persist in these waters because of an absence of competition with an indigenous fishery. No other lakes in the Missouri River drainage within Glacier National Park contain cutthroat trout, although introduced rainbow trout are present in several waters.

Summary and Conclusions

Status of the westslope cutthroat trout in Glacier National Park has been profoundly influenced by human activities during the past 75 years. Despite transboundary invasions and repeated introductions of nonnative salmonids, secure populations of westslope cutthroat trout continue to inhabit many lakes where they were historically present.

Available evidence suggests that 16 geographically discrete populations of westslope cutthroat trout originally inhabited 19 park lakes in the North Fork and Middle Fork drainages of the Flathead River. These waters encompass approximately 5,551 hectares. Inclusion of several small lakes suspected of harboring transplanted populations of native trout increases the area to 5,591 hectares.

Although substantial declines in populations of westslope cutthroat trout have been observed in several large westslope lakes, the species has not disappeared from any lake where it was historically present. Arrow Lake is the only lake in Glacier National Park where introgression has perturbed an indigenous cutthroat trout population.

The number of lakes inhabited by native cutthroat trout has not been reduced, but the presence of nonnative salmonids in Kintla, Bowman, Logging, Harrison, and McDonald lakes may pose a threat to the long-term security of the westslope cutthroat trout throughout a substantial portion of its historic range within the park. Lacustrine environs disturbed by the presence of nonnative salmonids amount to 84% of the area originally inhabited by westslope cutthroat trout in the North and Middle Fork drainages in Glacier National Park. Introduced Yellowstone cutthroat trout and hybrid populations of westslope cutthroat trout × Yellowstone cutthroat trout occur in five formerly fishless lakes encompassing 181 hectares in the Flathead River drainage of Glacier National Park.

The historic distribution of westslope cutthroat trout in the South Saskatchewan River basin remains obscure. A few remnant populations may still exist in some protected headwaters reaches, but to date none have been located.

Westslope cutthroat trout are not present in the upper Missouri River drainage within the park boundary. Isolated populations of Yellowstone cutthroat trout occur in three park lakes in the Missouri River drainage.

Although fish are no longer stocked in Glacier National Park, disturbances to the aquatic ecosystem resulting from past introduction of nonnative species may be irreversible. Some waters remain susceptible to invasions by nonnative migratory species, including several lakes which presently contain undisturbed native fisheries. The potential for such intrusions is underscored by the recent discovery of lake trout in Logging Lake.

Ecological effects of nonnative salmonids on westslope cutthroat trout are presently unclear. Competition for food between introduced species and native trout and predation by lake trout could be causes of increased natural mortality among native westslope cutthroat trout. Since most nonnative species have been present for less than 50 years, it may be too soon for an accurate assessment of the long-range consequences.

References

Butler, R. L. 1975. Some thoughts on the effects of stocking hatchery trout on wild trout populations. Pages 83–86 *in* W. King, editor. Wild trout management symposium. Trout Unlimited, Vienna, Virginia.

Crossman, E. J. 1984. Introduction of exotic fishes into Canada. Pages 78–101 *in* W. R. Courtenay, Jr., and J. R. Stauffer, editors. Management and biology of exotic fishes. Johns Hopkins University Press, Baltimore, Maryland.

Hanzel, D. A. 1959. The distribution of cutthroat trout *(Salmo clarki)* in Montana. Proceedings of the Montana Academy of Science 19:32–71.

Leathe, S. A., and P. J. Graham. 1982. Flathead Lake fish food habits study. Montana Department of Fish, Wildlife, and Parks, EPA contract, Kalispell.

Marnell, L. F. 1985. Aquatic resources policy and fisheries management in Glacier National Park. Proceedings of the Annual Conference Western Association of Fish and Wildlife Agencies 65:212–217.

Marnell, L. F. 1986. Impacts of hatchery stocks on wild fish populations. Pages 339–347 *in* R. H. Stroud, editor. Fish culture in fisheries management. American Fisheries Society, Fish Culture Section and Fisheries Management Section, Bethesda, Maryland.

Marnell, L. F., R. J. Behnke, and F. W. Allendorf. 1987. Genetic identification of cutthroat trout *(Salmo clarki)* in Glacier National Park, Montana. Canadian Journal of Fisheries and Aquatic Sciences 44:1830–1839.

Morton, W. M. 1968. Fisheries review report number 3: Many Glacier subdistrict. Glacier National Park, National Park Service Archives, West Glacier, Montana.

Moyle, P. B., H. Li, and B. A. Barton. 1986. The Frankenstein effect: impact of introduced fishes on native fishes in North America. Pages 415–426 *in* R. H. Stroud, editor. Fish culture in fisheries management. American Fisheries Society, Fish Culture Section and Fisheries Management Section, Bethesda, Maryland.

Schultz, L. P. 1941. Fishes of Glacier National Park, Montana. National Park Service, Glacier National Park, Conservation Bulletin 22, West Glacier, Montana.

Taylor, J. N., W. R. Courtenay, Jr., and J. A. McCann. 1984. Known impacts of exotic fishes in the continental United States. Pages 322–373 *in* W. R. Courtenay, Jr., and J. R. Stauffer, editors. Management and biology of exotic fishes. Johns Hopkins University Press, Baltimore, Maryland.

Willock, T. A. 1969. Distributional list of fishes in the Missouri River drainage of Canada. Journal of the Fisheries Research Board of Canada 26:1439–1449.

Note Added in Proof

In summer 1988, when this book was going to press, lake trout have been reported caught from Lower Quartz Lake. Although not confirmed, the author considers the identification reliable. If the presence of lake trout is verified, it will reduce by one the number of lakes in Glacier National Park that contain secure native cutthroat trout populations.

American Fisheries Society Symposium 4:71–74, 1988

Greenback Cutthroat Trout Recovery Program: Management Overview

ROBERT J. STUBER
U.S. Forest Service, Laconia, New Hampshire 03247, USA

BRUCE D. ROSENLUND
U.S. Fish and Wildlife Service, Golden, Colorado 80401, USA

JAMES R. BENNETT
Colorado Division of Wildlife, Denver, Colorado 80216, USA

Abstract.—The greenback cutthroat trout *Salmo clarki stomias* was the native trout species in the South Platte and Arkansas river drainages in Colorado. It was extirpated from much of its native range by the 1930s and is currently classified as a threatened species. Implementation of the current (1983) recovery plan is an interagency effort between the U.S. Fish and Wildlife Service, the Colorado Division of Wildlife, the U.S. National Park Service, the U.S. Forest Service, and the U.S. Bureau of Land Management. Activities have focused on the maintenance of remanent populations and the establishment of new populations throughout the historic range. This subspecies will no longer be considered threatened when 20 stable populations (i.e., those with multiple age-classes and natural reproduction) have been documented. Five "pure" remanent populations have been documented, and 17 reintroductions have taken place. Currently, there are seven stable populations. Fishing closures, stocking, better land-use practices, and habitat improvement have been utilized to maintain and enhance all populations. Three reintroduced populations have been opened to catch-and-release angling. Biological and angler responses have been favorable, and seven more areas may be opened to limited angling over the next 5 years. Although the recovery program has experienced a great deal of biological success, societal success has been less than desired, especially in regards to proposed future reintroductions. Public concerns have focused on the use of fish toxicants for nonnative trout removal, loss of existing angling opportunities, and subsequent needs for protective angling restrictions. It is apparent that a strong program of public education and involvement will be necessary to successfully carry out the mandate of the U.S. Endangered Species Act.

The greenback cutthroat trout *Salmo clarki stomias* was the only trout native to the South Platte and Arkansas river basins in Colorado except for the now-extinct yellowfin cutthroat trout in Twin Lakes. Biology and ecology of the greenback cutthroat trout have been described by Behnke and Zarn (1976) and Behnke (1979). This native cutthroat trout subspecies was extirpated from much of its historic range by the 1930s. Primary reasons for the decline were habitat degradation and the widescale introduction of nonnative trout species, which led to competition or hybridization, or both.

The greenback cutthroat trout is presently classified as threatened under the U.S. Endangered Species Act, which is the legal mandate for protection of threatened and endangered species in the USA. The law directs the Secretary of the Interior, in cooperation with other federal and state agencies, to develop recovery plans for threatened and endangered species that will allow such species to be upgraded from their current classification.

Implementation of the current (1983) Greenback Cutthroat Trout Recovery Plan is an interagency effort between the U.S. Fish and Wildlife Service, the Colorado Division of Wildlife, the U.S. Forest Service, U.S. National Park Service, and the U.S. Bureau of Land Management. The goal of this recovery plan is the removal of the greenback cutthroat trout from the threatened category by the year 2000. It will no longer be considered threatened when 20 stable populations (i.e., those with multiple-age classes and natural reproduction) have been documented throughout the subspecies' historic range (USFWS 1983). The purpose of this paper is to present an overview of the activities that are presently being undertaken to implement this recovery plan.

Maintenance of Historic Populations

To date, surveys for remanent populations have led to the documentation of five genetically "pure" populations (Binns 1977), all of which occupy isolated headwater reaches of streams or alpine lakes. These populations are monitored by the cooperators in the recovery plan to determine trends in both population and habitat characteristics. Information derived from monitoring these pure populations is used to make future management recommendations to maintain and enhance existing and future populations. For instance, it was determined from monitoring greenback cutthroat trout in Como Creek that this subspecies exhibits a preference for pool areas even though riffles and runs make up the majority of its overall habitat (U.S. Forest Service and Colorado Division of Wildlife, unpublished data). Based on these data, habitat enhancement was conducted by installing check dams constructed of logs to create scour pools. Future monitoring will determine the actual population response to this enhancement work.

There are a number of other management strategies being employed to maintain remanent populations. The stocking of nonnative trout into habitats occupied by greenback cutthroat trout is strongly discouraged. Also, the existence of these remanent populations is not highly publicized in order to discourage any illegal harvest. Finally, the governing agencies for those areas are encouraged to employ appropriate land management practices so that existing habitats are not degraded.

Establishing New Populations

Much of the recovery effort to date has focused on the reintroduction of greenback cutthroat trout into suitable habitat throughout its historic range. The first step in this process is the identification of potential reintroduction sites. Criteria that are utilized in this identification phase are good quality trout habitat (i.e., habitat capable of supporting a wild trout standing crop of at least 22 kg/hectare); low to moderate fishing use; presence of a barrier or potential barrier site; and good reclamation potential (i.e., few beaver ponds, not many tributaries). Preference is also given to those potential reintroduction sites located in areas where the management emphasis is on native ecosystems, such as designated wilderness areas or national parks.

Once a reintroduction site has been selected, the National Environmental Policy Act (NEPA) documentation takes place. This is a detailed process which includes public involvement, a description of the affected environment, and a full disclosure of all environmental consequences associated with the implementation of the proposed reintroduction. Upon NEPA compliance, certain site preparations are necessary prior to the actual reintroduction. Because most of the selected reintroduction sites have an existing nonnative trout population, two problems must then be dealt with: (1) removal of this nonnative trout population; and (2) prevention of repopulation by nonnative trout. In practice, the initial step is the location or creation of a barrier to prevent reinvasion by nonnative fish. Natural features such as waterfalls (1.5 – 3m in height) will serve as effective fish barriers. Where no such feature exists, a barrier can be constructed. Rosgen and Fittante (1986) have developed guidelines for migration barrier placement. Once a barrier is located or in place, the second step is to remove the nonnative trout from the stream or lake above this barrier. This is done through the application of a piscicide, either rotenone or antimycin. Both of these substances have been approved by the Environmental Protection Agency for controlled use in fishery management activities. Fish toxicant is limited to the designated area by the introduction of potassium permanganate, an oxidizing agent, immediately downstream of the fish barrier.

The macroinvertebrate forage base may be affected by the piscicide treatment, and lack of a suitable forage may lessen the chances for survival of reintroduced native trout (Rinne et al. 1981). However, it is thought that forage base recovers to pretreatment levels within 1–6 months following application and neutralization of the piscicide. Most reclamations take place in late summer, and the actual reintroductions of greenback cutthroat trout take place the following year to ensure that a suitable forage base will be present.

Most of the reintroductions have utilized a fry-stocking program. Fry are scatter-planted throughout the receiving body of water for three or four consecutive years in an attempt to simulate natural reproduction and to establish a multiple-age-class population. Most of the reintroduction sites are relatively inaccessible, and a variety of transport mechanisms such as helicopters, llamas, and backpacking must be employed for the stocking. Fry are obtained from the U.S. Fish and Wildlife Service Fish Cultural Development Center in Bozeman, Montana (Dwyer and Rosenlund 1988, this volume). Brood stock at the Bozeman hatchery was developed from several wild populations. In addition, milt from wild males is collected each year and sent to the Bozeman facility to fertilize the eggs, ensuring that the genetic diversity of this hatchery source is maintained.

Reintroductions are monitored to determine if and when a self-sustaining population has been established. Growth and survival of reintroduced greenback cutthroat trout are determined, and natural reproduction is documented. Monitoring also provides information to help make recommendations regarding other existing or proposed land uses.

Habitat enhancement has been undertaken at several sites in an effort to improve the chances of success for the reintroduction. Most of this work has focused on creating additional pool habitat through the installation of log check dams. Estimated carrying capacity for trout has increased by 110% (average increase of 34 kg/hectare) following the installation of such structures in small- and medium-size Colorado streams (Stuber 1986). Such increases probably are the result of enhanced overwinter habitat, because the number of holding areas (pools) often limit the carring capacity of Rocky Mountain streams during periods of low flow in winter. Pools are important as refuge areas during winter (Raleigh et al. 1984) when trout move to cover in pool areas to avoid physical damage from ice scouring (Hartman 1965; Chapman and Bjornn 1969) and to conserve energy (Everest and Chapman 1972). Therefore, formation of scour pools through the installation of log check dams should enhance the survival of the reintroduced greenback cutthroat trout populations.

Current Status of Recovery Program

Five pure remanent populations of greenback cutthroat trout have been located; they occupy four streams (13 km) and one lake (5 hectares). Greenback cutthroat trout have been reintroduced into 12 streams (64 km) and 5 lakes (20 hectares) in Colorado. This amounts to less than 5% of the total coldwater habitat available throughout the subspecies' historic range. Most of these 17 introductions have taken place in the Arapaho–Roosevelt and Pike–San Isabel national forests and in Rocky Mountain National Park, and most were initiated within the last 5 years (1983–1987). Currently 7 of these 22 populations (remanent and reintroductions) are categorized as stable.

All remanent populations are closed to angling to protect the greenback cutthroat trout. It is intended to maintain these protective closures because these areas are managed as refugia. A reintroduced population is initially closed to angling until a self-sustaining, stable population is established, at which time, the population may be managed with a protective catch-and-release regulation. Colorado statute prohibits the "taking" of a threatened species; however, the Division of Wildlife does not consider that catch-and-release constitutes taking. Fishing for greenback cutthroat trout was initiated on an experimental basis at Hidden Valley Creek within Rocky Mountain National Park in 1982. This was done in an effort to control the number of nonnative brook trout *Salvelinus fontinalis*, which are still in that particular system. Brook trout are managed under a catch-and-keep regulation. The fishery has been quite popular, and it was estimated, from the 1982-1983 creel census, that 2,200 angler-hours were expended and 1,760 greenback cutthroat trout were caught and released. Two more reintroduced populations within Rocky Mountain National Park, Ouzel Lake and Fern Lake, were opened to fishing in 1986. It is anticipated that as many as seven additional waters in both the Arapaho–Roosevelt and Pike–San Isabel national forests and Rocky Mountain National Park may be opened to angling over the next 5-year period (1986-1991) as reintroduced populations become stable.

Although the recovery program has experienced considerable biological success, its societal success has been less than desired. There has been a lack of public support, especially in regards to proposed future reintroductions. Public concern has focused mainly on three topics: (1) the use of fish toxicants; (2) loss of existing fishing opportunities; and, (3) future protective angling restrictions.

The use of fish toxicants to remove nonnative trout prior to the reintroduction of greenback cutthroat trout has generated a great deal of public concern. Some of this concern is philosophical in nature and is related to the need to kill one species of trout to restore another. Also, some people feel it is arbitrary for resource management agencies to allow the killing of trout (harvest) in one location and to adamantly protect them in another. Other concerns are safety oriented, such as whether the toxicant is actually contained within the designated treatment area and the potential contamination of groundwater supplies (downstream wells) and disruption and contamination of the aquatic based food chain.

The temporary (3–6 year) loss of existing fishing opportunities and future protective fishing restrictions associated with a reintroduction have also generated some public concern. Adherence to the reintroduction criteria of low-to-moderate existing angling use has generally been followed. There is, however, a perceived loss of a much larger magnitude by some members of the public. Objections regarding future angling restrictions are more philosophical in nature (e.g., catch-and-keep versus catch-and-release).

It has become quite apparent that a strong information and education program will be necessary to successfully carry out the mandate of the Endangered Species Act. Resource managers involved in the program need to sell it to the public, especially by explaining the overall philosophy of the law and its application to restoration activities, and to keep the public informed of the specific management activities associated with this restoration program. Education also is needed about the use of fish toxicants (e.g., necessity of use, application procedures, and actual hazards and safety precautions). Anglers need to be made aware that greenback cutthroat trout reintroductions will ultimately offer them a unique recreational opportunity to fish for a rare species of trout, and that monitoring programs of existing and future greenback cutthroat trout fisheries will yield information relative to proposed future reintroductions and projected unique fishing opportunities. The relatively small amount of habitat involved with the restoration program should be emphasized to allay the fears of those who perceive a loss in angling opportunities. Above all, public involvement in the restoration program will be necessary for it to be successful. This involvement should include information exchange and, whenever possible, active participation in the actual restoration projects. There are trade-offs involved with the greenback cutthroat trout recovery program, but a public awareness of the long-term benefits of restoring this native species compared to the short-term impacts associated with the implementation of the program should improve its acceptance and support.

References

Behnke, R. J. 1979. Monograph of the native trouts of the genus *Salmo* of western North America. Report to U.S. Fish and Wildlife Service, Denver, Colorado.

Behnke, R. J., and M. Zarn. 1976. Biology and management of threatened and endangered western trouts. U.S. Forest Service General Technical Report RM-28.

Binns, N A. 1977. Present status of indigenous populations of cutthroat trout *Salmo clarki* in southwestern Wyoming. Wyoming Game and Fish Department Technical Bulletin 5.

Chapman, D. W., and T. C. Bjornn. 1969. Distribution of salmonids in streams, with special reference to food and feedings. Pages 153–176 *in* N. G. Northcote, editor. Salmon and trout in streams. H. R. MacMillan Lectures in Fisheries, University of British Columbia, Vancouver.

Dwyer, W. P., and B. D. Rosenlund. 1988. Role of fish culture in the reestablishment of greenback cutthroat trout. American Fisheries Society Symposium 4:75–80.

Everest, F. H., and D. W. Chapman. 1972. Habitat selection and spatial interactions by juvenile chinook salmon and steelhead trout in two Idaho streams. Journal Fisheries Research Board Canada 29:91–100.

Hartman, G. F. 1965. The role of behavior in the ecology and interaction of underyearling coho salmon (*Oncorhynchus kisutch*) and steelhead trout (*Salmo gairdneri*). Journal of Fisheries Research Board Canada 22:1035–1081.

Raleigh, R. F., T. Hickman, R. C. Solomon, and P. C. Nelson. 1984. Habitat suitability information: rainbow trout. U.S. Fish and Wildlife Service Biological Services Program FWS/OBS-82 10.60.

Rinne, J. H., W. L. Minckley, and J. H. Hansen. 1981. Chemical treatment of Ord Creek, Apache County, Arizona, to re-establish Arizona trout. Journal of Arizona–Nevada Academy of Science 16:74–78.

Rosgen, D. and B. L. Fittante. 1986. Fish habitat structures—a selection guide using stream classification. Pages 163–179 *in* J. G. Miller, J. A. Arway, and R. F. Carline, editors. Proceedings of the fifth trout stream habitat improvement workshop. Pennsylvania Fish Commission, Harrisburg.

Stuber, R. J. 1986. Stream habitat improvement evaluation: an alternative approach. Pages 153–161 *in* J. C. Miller, J. A. Arway, and R. F. Carline, editors. Proceedings of the fifth trout stream habitat improvement workshop. Pennsylvania Fish Commission, Harrisburg.

USFWS (U.S. Fish and Wildlife Service). 1983. Greenback cutthroat trout recovery plan. Greenback Cutthroat Trout Recovery Team, Denver, Colorado.

American Fisheries Society Symposium 4:75–80, 1988

Role of Fish Culture in the Reestablishment of Greenback Cutthroat Trout

WILLIAM P. DWYER

U.S. Fish and Wildlife Service, Fish Technology Center
4050 Bridger Canyon Road, Bozeman, Montana 59715, USA

BRUCE D. ROSENLUND

U.S. Fish and Wildlife Service, Colorado Fish and Wildlife Assistance Office
730 Simms, Suite 292, Golden, Colorado 80401, USA

Abstract.—The greenback cutthroat trout *Salmo clarki stomias* is indigenous to the South Platte and Arkansas river drainages on the east slope of the Colorado Rockies. Human development, habitat destruction, and the introduction of nonnative species have caused the subspecies to be eliminated from most of its native range. A recovery program was initiated and, in 1977, 66 fish were removed from Como Creek (Colorado), one of the two remanent populations known to exist at that time, and transferred to the U.S. Fish and Wildlife Service Fish Technology Center in Bozeman, Montana. These fish were used to develop a captive brood-stock population. From 1979 through 1986 over 160,000 fry were returned to Colorado for reintroduction into the subspecies' native habitat. Sperm from wild males is used to help maintain the wild gene pool at Bozeman. Electrophoretic and asymmetry analyses are being used to monitor the genetic integrity of the population. One captive and two semiwild brood-stock populations have been established from the native greenback cutthroat trout of the Arkansas River drainage (Colorado). Saratoga National Fish Hatchery (Wyoming) began brood-stock production in 1987. Efforts to reestablish the greenback cutthroat trout have been successful. To date, 81 km of streams and 44 hectares of lakes and ponds have been repopulated with fry from the captive brood-stock program. Three of the restoration projects have been opened to catch-and-release fishing.

Historical Background

The greenback cutthroat trout *Salmo clarki stomias* is indigenous to the Colorado headwaters of the South Platte and Arkansas river drainages. This subspecies originated several thousand years ago when a headwater tributary in the Colorado River basin became connected to a headwater stream in the South Platte River drainage. This connection allowed the Colorado River cutthroat trout *Salmo clarki pleuriticus* to cross to the east slope of the Continental Divide and establish populations from which the greenback cutthroat trout evolved and later moved into the Arkansas River drainage. Reports indicate that greenback cutthroat trout were once abundant in all mountain and foothill streams along the Front Range of the Colorado Rocky Mountains. However, by the end of the 19th century their abundance had drastically declined, and they were replaced by repeated stockings of brown trout *Salmo trutta* and rainbow trout *Salmo gairdneri* (Behnke 1979).

The greenback cutthroat trout, like many inland subspecies of cutthroat trout, does not coexist well with nonnative trout species. The stocking of brook trout *Salvelinus fontinalis* in higher-elevation streams resulted in the loss of many greenback cutthroat trout populations because greenback cutthroat trout are not able to survive in competition with this species (Fausch and Cummings 1986).

In 1889, David Star Jordan (Jordan 1891) was sent by the U.S. Commissioner of Fish and Fisheries to survey the streams of Colorado and Utah. The primary purpose was to determine the present stock of food fishes in these waters and the suitability of the waters for the introduction of other desirable species. The secondary objective was to catalog the native fishes of each stream, "whether food-fishes or not," and to better understand the geographical distribution of the fishes.

On this trip Jordan noted that deterioration of water quality due to placer mining and stamp mills had rendered waters of otherwise clear streams yellow or red with clay and had made the streams uninhabitable for trout. Many parts of the Arkansas and Grand rivers had been ruined as trout stream by mining operations.

Jordan (1891) also reported that irrigation and habitat loss had taken a substantial toll.

> In the progress of settlement of the valleys of Colorado the streams have become more and more largely used for irrigation. Below the mouths of the canons [sic] dam after dam and ditch after ditch turn off the water. In summer the beds of even large rivers (as the Rio Grande) are left wholly dry, all the water being turned into these ditches. . . .

He goes on to say,

> Great numbers of trout, in many cases thousands of them, pass into these irrigating ditches and are left to perish in the fields. The destruction of trout by this agency is far greater than that due to all others combined, and it is going on in almost every irrigating ditch in Colorado.

The greenback cutthroat trout was extirpated from most of its natural range by the early 20th century, and Greene (1937) considered the subspecies to be extinct. However, when the U.S. Endangered Species Act was passed in 1973, small native populations of the rare greenback cutthroat trout were known to exist in Como Creek and the South Fork of the Cache La Poudre River (tributaries in the South Platte River drainage) in Colorado. These populations inhabited a combined stream length of 6.4 km and numbered about 2,000 individuals. Meristic characteristics of fish sampled from these populations conformed to characteristics of type specimens collected in 1856. Due to the small size of these populations the subspecies was listed as "endangered" in 1973. Over the next few years three more native populations were discovered, and in 1978, the greenback cutthroat trout was reclassified as "threatened" in order to facilitate recovery efforts.

The goal of the U.S. Fish and Wildlife Service's (USFWS) Greenback Cutthroat Trout Recovery Plan (USFWS 1983) is the recovery of this subspecies so that it can be removed from the threatened species list. This subspecies will be considered recovered when 20 stable type-A (i.e., genetically pure) greenback cutthroat trout populations are documented within its historic range. Stuber et al. (1988, this volume) discuss the objectives of the recovery plan in greater detail.

In order to reduce the amount of time required to accomplish the goal and reestablish fishable populations, the Greenback Cutthroat Trout Recovery Team requested that captive brood stocks be established to produce fry for stocking restoration projects. The first captive brood stock came from the South Platte–Como Creek population and was held at the USFWS, Fish Technology Center (FTC), Bozeman, Montana. Because attempts to establish captive greenback cutthroat trout brood stocks had not been successful in 1889 or 1957, semiwild brood stocks were also established with fish from the Arkansas River drainage at a private ranch (Mc Alpine Pond) in Huerfano County and at Fort Carson Military reservation (Lytle Pond) both in Colorado.

Captive Broodstock

South Platte River Drainage

The first step in establishing a captive greenback cutthroat trout brood stock at the Bozeman FTC was to obtain a disease history of the Como Creek populations. The limited numbers of Como Creek fish precluded killing 60 of them for the normal disease certification procedure. To assist the recovery team, the USFWS Fish Disease Control Center examined fecal material and ovarian and seminal fluid from 78 pre- and postspawning Como Creek greenback cutthroat trout. None of the control center's listed pathogens were identified in the samples.

Based upon the available disease data, authorization was granted, and in September 1977, 64 fish (mean weight, 26.6 g) were transferred to the Bozeman FTC. These fish were segregated from other salmonids, and mortalities were monitored. The fish were divided into two groups. The 32 smaller ones (mean weight, 10.2 g) were placed in an indoor tank supplied with 10°C spring water to enhance growth. The remaining 32 fish (mean weight, 43 g) were raised in an outdoor raceway in creek water. Smith et al. (1983) found that the percent of westslope cutthroat trout *Salmo clarki lewisi* eggs to reach eye-up improved dramatically if the fish overwintered in creek water rather than constant 10°C spring water.

Problems were encountered immediately with the captive brood stock. The fish refused commercial diets as well as some natural foods. A high-energy palatable diet consisting of beef liver, salt, cod liver oil, unflavored gelatin (binder), and commercial feed was developed. These ingredients were thoroughly ground, and vermiform portions were produced with a 100 mL syringe.

Because the fish were small, we thought that it would be less stressful if they were allowed to spawn naturally in 1978. Spawning boxes were constructed with a large-mesh screen bottom. The screen was covered with gravel and small stones that were larger than the mesh. Fish were expected to spawn in the gravel, and eggs could then be removed from the bottom box for counting and incubation, a technique that had been used successfully with brook and rainbow trout in toxicological research.

Asynchronous maturation of males and females occurred the first spawning season the fish were in captivity. Similar behavior was noted with captive greenback cutthroat trout at the Leadville (Colorado) National Fish Hatchery (NFH). Our males produced milt in April and May, but the females matured in July and August. In an attempt to obtain eggs, fish were artificially spawned. Five females were spawned, and 1,005 eggs (mean ± SD, 201 ± 67 eggs/female) were obtained. However, most eggs were poorly embryonated, and only 13 offspring survived.

In 1979, spawning was successful because the sexes matured synchronously; this was the first time greenback cutthroat trout had been spawned successfully in captivity. Four females were spawned, and 81% of the eggs reached the eyed stage. Five hundred fry began feeding, but mortalities increased in October due to a severe bacterial enteritis. Fry were treated successfully with sulfa for 10 d, and 268 young survived. Growth rate through August 1980, was 0.051 cm/d. Mean weight and length were 91 g and 21 cm, respectively.

In December 1979, 15 of the original 64 brood-stock fish remained. These fish were at least 5 years old, and some were considerably older. Advanced age and spawning stress apparently increased their susceptibility to pathogens. Losses occurred mainly in the males, which failed to respond to treatment for *Saprolegnia* infections. In 1980, only 8 of the original Como Creek stock and 12 age-2 fish from the 1978 spawning remained. The age-2 fish averaged 30 cm in length and were slightly larger than the original Como Creek stock. Three of the 1978 fish were ripe in 1980, but the eggs were nonviable. It was assumed that this condition occurred because these fish had been held at 13°C in order to attain better growth rates.

Seven of the original Como Creek fish were spawned during June and July 1980. Mean (SD) number of eggs per female was 529(159). Survival to swim-up was poor (48%) due to a low hatching success. The surviving young (N = 900) grew 9.64 cm from October 1980 through June 1981, or 0.037 cm/d. Temperature was held at 12°C until February when it was decreased to 10°C in order to reduce growth and provide the size of fish requested for stocking.

On June 26, 1981, 865 yearling greenback cutthroat trout, averaging 15.5 cm and 36 g, were successfully returned to Colorado. Bear Lake in Rocky Mountain National Park received 432 fish. The inlet and outlet streams of the lake had recently been improved to create spawning habitat. The remaining fish were given to the Colorado Division of Wildlife for stocking into Williams Gulch and Hourglass Creek at a later date.

The 1981 spawning at the Bozeman FTC was the most successful one up to that time. Spawning began in May and continued through July. Eggs were collected from 101 females—90 females were from age-2 brood stock and the remaining 11 were from the original Como Creek stock and the age-3 brood-stock fish. Age-2 fish produced 35,000 eggs (mean, 389 eggs/female) of which 60% reached the eyed stage of development. The age-3 fish produced 11 spawns, totaling 6,065 eggs (551 eggs/female), of which 80% reached the eyed stage. Mean weights were 254 g and 357 g for age-2 and age-3 fish, respectively.

Sixty-nine percent or 16,570 fry survived from the eyed stage. They were transported by air to Colorado on September 15 for introduction into Ouzel Lake and its drainage. Fry were placed into plastic bags at a density of 120 g/L. Ice was added to lower the temperature to 2–4°C. Bags were inflated with oxygen, sealed, and placed in insulated ice chests. The fish arrived in excellent condition and were immediately taken to Rocky Mountain National Park where a helicopter was waiting to take them to the lake.

Since 1981, sufficient numbers of greenback cutthroat trout fry have been raised each year to meet the needs of the recovery team. To date, 160,000 fry have been raised and transported to Colorado for introduction into reclaimed native waters.

Milt has been collected from native populations for use at Bozeman FTC since 1982. This has been done in order to help maintain the gene pool, to compensate for the loss of males at the hatchery, and to alleviate problems due to asynchronous maturation. Como Creek and Hidden Valley Creek (a restoration project) greenback cutthroat trout spawn in late May to June and have been crossed with the Bozeman brood stock. Sperm from wild males is expressed into a 15 mL tube, which is kept on ice and shipped by air freight. The sperm is used within 24 h of the time it was taken.

The second attempt to establish a captive South Platte River brood stock involved the Saratoga NFH, Wyoming, and the natural population of greenback cutthroat trout in the South Fork of the

Cache La Poudre River (Poudre River) in Colorado. This population is small (approximately 640 fish), does not spawn until July during most years, and does not appear to successfully reproduce each year. Following a disease examination based upon fecal material and ovarian and seminal fluids as previously described for Como Creek fish, eggs were collected in 1984 and 1985. The eggs were shipped to the Saratoga NFH and held in an isolation unit. Eggs collected in 1984 did not survive, but 47% of the eggs collected in 1985 survived to swim-up. None of those young accepted feed, and all died.

Eggs from the Poudre River population required much less time to develop and hatch than did those of the greenback cutthroat trout from the Arkansas drainage's Cascade Creek. At 8°C, eggs from the Poudre River fish required only 16 d to reach the eyed stage and 32 d to hatch compared to 29 d to the eyed stage and 39 d to hatch for Cascade Creek fish (J. Hammer, Saratoga NFH, personal communication).

To increase the genetic diversity of the Como Creek brood stock at Bozeman FTC, milt was collected from the Poudre River fish in 1986. This was used to fertilize ova from the Bozeman Como Creek brood-stock line. Progeny from this cross are being held for future brood stock.

The Hunters Creek population, within the Rocky Mountain National Park, was discovered in June 1985. The fish were confirmed as type-A greenback cutthroat trout (R. J. Behnke, Colorado State University, personal communication). The number of fish over 150 mm in the creek was estimated at 1,080 in 1985, and they inhabited about 2 km of the stream.

During June 1986 and 1987, milt from Hunters Creek greenback cutthroat trout was collected to use with the Como Creek brood stock at Bozeman. This will improve the genetic diversity of the future Bozeman brood stock and, therefore, the quality of the fry used in the restoration projects.

Arkansas River Drainage

The only pure native population of Arkansas River greenback cutthroat trout known to exist was discovered in Cascade Creek in 1977. This population, estimated at 620 fish, is located in an isolated area. As an alternative to captive hatchery programs, semiwild brood stocks were established from this population in 1980 and 1981. Adult and subadult Cascade Creek fish were introduced into McAlpine Pond on the McAlpine Ranch and at Lytle Pond on Fort Carson.

No eggs were collected from McAlpine Pond in 1981 or 1982. A few eggs were collected in 1983, but none survived. In June 1984, five females produced 5,556 eggs, which were fertilized with sperm from six males. Eggs were shipped to the Saratoga NFH where they were held in an isolation unit. Survival of these eggs was excellent; 98% survived to eye-up, and 46% survived to 93 mm in length, at which time they were stocked into management areas.

Lytle Pond, constructed by the U.S. Army, is fed by a natural spring; it was planned to be a refugium for Arkansas River drainage fish at Fort Carson. In 1981, 40 Cascade Creek greenback cutthroat trout were stocked in the pond to start a semiwild brood stock. None of these trout exceeded 130 g. Growth of these fish was excellent; by 1984, 2.0 kg fish were found. In early May 1984, these fish spawned; 1,743 green eggs from that spawn were sent to the Saratoga NFH, and 2,280 eggs were incubated at the USFWS Laboratory in Denver, Colorado. Eye-up of these eggs was 56 and 51%, respectively. The 1,100 fry surviving from eggs incubated at the Denver laboratory were stocked back into Lytle Pond in June 1984. The survival and growth of these fry were excellent, which allowed some of the fish to be moved to new sites.

By 1985, most of the original stock of Cascade Creek fish appeared to be gone from Lytle Pond. However, progeny of the original fish had established a successful spawning population. To improve genetic diversity, 55 young greenback cutthroat trout from the McAlpine population were marked and stocked into Lytle Pond in 1985. These marked fish have been observed in the 1987 spawning run.

Since 1984, Lytle Pond fish have been allowed to spawn naturally; spawning begins about April 20, and the number of spawners has increased each year. Fry are collected in October in minnow traps and are used to stock restoration projects on Fort Carson and on national forest lands.

Eggs from the two semiwild brood stocks that were sent to Saratoga NFH in 1984 were used either to develop a captive brood stock or were reared for fish restoration projects. Following the 1985 stocking, future brood stock at Saratoga NFH consisted of 250 Lytle and 250 McAlpine greenback cutthroat trout. These fish spawned successfully for the first time in April 1987. (J. Hammer, Saratoga NFH, personal communication).

Greenback Cutthroat Trout Stocking

To insure the best genetic mix and provide for several year classes of greenback cutthroat trout in restoration projects, each project was scheduled to receive young-of-the-year fish (30–43 mm in length) for three consecutive years. The use of wild milt each year to fertilize hatchery-produced ova should enhance the genetic diversity of the fish used for restoration.

Initial stocking rates were high to compensate for expected losses due to stress during transportation. Although lake stocking never exceeded 1,800 fish/hectare per year, this rate was found to be too high for lakes near 3,000 m in elevation. Stocking Fern Lake in Rocky Mountain National Park at an annual rate of 1,093 fish/hectare produced 220-mm cutthroat trout within two growing seasons; however, five growing seasons were necessary to produce the same size fish in Ouzel Lake, which had annual stocking rates of 1,728 fish/hectare.

Genetic Evaluation

Deformities, such as short opercules, bent or short peduncles, and bent jaws, were observed in 2–5% of the type-A greenback cutthroat trout. These occurred most frequently in populations founded by direct or indirect transfers from Como Creek.

To prevent the loss of variation by genetic drift and founder events within the Bozeman brood stock, milt from Hidden Valley and Como Creek populations has been used since 1982. With the use of milt from wild fish and the repeated stocking of restoration areas for 3 years, it was anticipated that problems associated with the loss of genetic variation would be minimized in restoration projects.

To determine levels of genetic variation within and the extent of genetic differences between the type-A greenback cutthroat trout populations, electrophoretic and asymmetry analyses of four populations were completed in 1986. Samples were collected from the Bozeman and Saratoga brood stocks and the Hidden Valley and Ouzel Lake restoration projects. The brood stocks represented the South Platte (Bozeman) and Arkansas (Saratoga) drainages. The Hidden Valley restoration project was founded from 84 adults and subadults transferred directly from Como Creek, and the Ouzel Lake restoration project was founded by a 3-year stocking program of over 28,000 young-of-the-year fish from Bozeman FTC.

Results of the electrophoretic analyses showed statistically significant differences in allele frequencies among the Bozeman, Hidden Valley, and Ouzel Lake populations. Since these three sample populations originated from the Como Creek stock, the presence of genetic differences suggests that these populations may not be genetically representative of the Como Creek population. The suspected cause of these differences is the reduced amount of genetic variation in the population caused by founder events and genetic drift. Asymmetry data provide support for this contention. Previous studies demonstrated that loss of genetic variation increases the number of bilateral meristic characters that have asymmetric counts (i.e., left count different from right count) in salmonid fishes (Leary et al. 1984, 1985). The three sample populations have exceptionally high levels of asymmetry (mean, 2.00 to 2.24 characters/individual out of 5 characters examined). The only other populations in which comparable levels of asymmetry have been observed are a highly inbred group of experimental rainbow trout and a hatchery population of westslope cutthroat trout. In contrast, the Saratoga NFH brood stock was founded from few individuals from the Arkansas drainage; asymmetry in that brood stock was 1.8 asymmetric characters/individual, which is within the range normally observed in salmonid fishes (R. F. Leary, University of Montana, personal communication).

The data presented by Leary et al. (1984) indicated that a negative correlation exists between the number of heterozygous loci and the number of asymmetric characters per individual. Due to the small number of type-A greenback trout remaining in the remanent native populations, it is possible that these populations may have low levels of genetic variation. If this is true, then a single population may not be a good sole source of genetic material for brood stocks. It would be better to use multiple sources. There is reservation about killing 25 fish from the native populations to obtain electrophoretic samples and test for heterozygosity. However, because a correlation exists between heterozygosity and asymmetry, changes in heterozygosity may be monitored with the asymmetry index. There is a collection of preserved native greenback cutthroat trout that could be analyzed for asymmetry for comparisons with the asymmetry of present brood stocks and fish in restoration projects.

Other work by Leary (personal communication) showed that cutthroat trout have less heterozygosity and more asymmetry than rainbow trout. Obvious deformities were observed in Snake River cutthroat trout from Jackson NFH. Analysis of the Jackson NFH Snake River brood stock, in 1986, showed that the brood stock had more asymmetry (10-12%) and lower levels of electrophoretic variation than any of the greenback cutthroat trout samples. These studies demonstrate potential genetic problems that can be encountered in founding and maintaining captive populations of cutthroat trout.

Conclusion

The Greenback Cutthroat Trout Recovery Team has established and maintained two captive and two semiwild greenback cutthroat trout brood stocks to produce fry for stocking restoration sites. Through interagency recovery efforts and the brood stock program, these fish have been introduced into 81 km of streams and 44 hectares of lakes and ponds. Presently three restoration sites are open to catch-and-release angling.

The use of hatcheries to perpetuate endangered species can be a valuable management method if used correctly, but many considerations should be studied before a captive brood-stock program is begun. Attitude of hatchery personnel is an important aspect. Hatchery managers are often more concerned about reaching a production commitment than about pampering a small group of wild fish that neither feed nor grow like domesticated rainbow trout.

Perhaps the most important consideration should be the genetic integrity of the brood stock. Measures must be taken to prevent genetic contamination from other hatchery fish; however, this precaution is not sufficient by itself. Selection factors differ between hatchery and natural environments. It is very important to periodically introduce original genetic material to prevent strong adaptation to hatchery conditions which would have deleterious effects when the fish are stocked. All the fish maintained for brood stock at Bozeman are progeny of Bozeman females and of males from wild populations.

The use of a captured population to produce progeny to repopulate a native area is not a solution unless habitats have been reclaimed from nonnative species or the cause for decline has been remedied. As stated by Rinne et al. (1986), "the role that hatcheries play in the recovery of rare fishes is only as good as the availability of suitable habitat in the wild for reintroduction."

References

Behnke, R. J. 1979. The native trouts of the genus *Salmo* of western North America. Report to U.S. Fish and Wildlife Service, Denver, Colorado.

Fausch, K. D. and T. R. Cummings. 1986. Effects of brook trout competition on threatened greenback cutthroat trout. Colorado State University, National Park Service contract CX-1200-4-A043, Ft. Collins.

Greene, W. D. 1937. Colorado trout. Colorado Museum of Natural History, Popular Series 2, Denver.

Jordan, D. S. 1891. Report of the explorations in Colorado and Utah during the summer of 1889, with an account of the fishes found in each of the river basins examined. U.S. Fish Commission Bulletin 9: 1–40.

Leary, R. F., F. W. Allendorf, and K. L. Knudsen. 1984. Superior development stability of enzyme heterozygotes in salmonid fishes. American Naturalist 124:540–551.

Leary, R. F., F. W. Allendorf, and K. L. Knudsen. 1985. Developmental instability as an indicator of reduced genetic variation in hatchery trout. Transactions of the American Fisheries Society 114:230–235.

Rinne, J. N., J. E. Johnson, B. L. Jensen, A. W. Roger, and R. Sorenson. 1986. The role of hatcheries in the management and recovery of threatened and endangered fishes. Pages 271–285 in R. H. Stroud, editor. Fish culture in fisheries management. American Fisheries Society, Fish Culture Section and Fisheries Management Section, Bethesda, Maryland.

Smith, C. E., W. P. Dwyer, and R. G. Piper. 1983. Effect of water temperature on egg quality of cutthroat trout. Progressive Fish-Culturist 45:176–178.

Stuber, R. J., B. D. Rosenlund, and J. R. Bennett. 1988. Greenback cutthroat trout recovery program: management overview. American Fisheries Society Symposium 4:70–74.

USFWS (U.S. Fish and Wildlife Service). 1983. Greenback cutthroat trout recovery plan. Greenback Cutthroat Trout Recovery Team, Denver, Colorado.

American Fisheries Society Symposium 4:81–89, 1988

Identification and Status of Colorado River Cutthroat Trout in Colorado

ANITA M. MARTINEZ

Colorado Division of Wildlife, 711 Independent Avenue, Grand Junction, Colorado 81505, USA

Abstract.—From 1978 to 1987, 135 streams and 25 lakes were surveyed in northwestern Colorado for the presence of pure populations of Colorado River cutthroat trout *Salmo clarki pleuriticus*. The subspecies was not found in 59 of these streams or in 11 of the lakes. Meristic characters and spotting pattern served as indicies of genetic purity for 1,404 specimens. Spotting pattern (variability in number, size, shape, and position of spots) was considered a valid character within a population but not between populations. One lake and 36 streams were monitored for strain purity, population stability, and available habitat. Fourteen of these 37 populations maintained previous purity ratings, 7 increased their ratings, and 12 lowered them. Additionally, three populations were replaced by brook trout *Salvelinus fontinalis,* and one disappeared without replacement. Eighteen streams and 3 lakes identified to date contain relatively pure populations of Colorado River cutthroat trout. This subspecies may be in jeopardy in Colorado. Additional research and management efforts to protect existing populations and to reclaim portions of the subspecies' historic range are warranted to preserve this cutthroat trout genotype.

The Colorado River cutthroat trout *Salmo clarki pleuriticus* was first described by Cope (1872) from specimens collected from the Green River, Fort Bridger, Wyoming. Cope, however, described a keel along the midline of the skull that actually had resulted from improper preservation and consequent dehydration (Behnke 1979). Cope (1872) also applied the name *pleuriticus* to cutthroat trout in the South Platte and Yellowstone rivers, and in Rio Grande and Bonneville basins. Jordan (1891) later assigned *pleuriticus* solely to Colorado River basin cutthroat trout. Because *pleuriticus* was the first name applied to Colorado River basin cutthroat trout it became the valid nomenclature.

The Colorado River cutthroat trout since has been described in detail (Wernsman 1973; Behnke and Zarn 1976; Johnson 1976; Binns 1977; Behnke 1979; Behnke and Benson 1980). Its coloration is renowned. A vague to brilliant horizontal red band runs along each brassy yellow side, its intensity varying with the fish's diet. Mature males often have crimson ventral regions during spawning season. The spotting pattern, although geographically variable, consists of large round black spots concentrated posteriorly and above the lateral line anteriorly. Typical Colorado River cutthroat trout have 170 to 200 or more lateral-series scales counted two rows above lateral line, 25–45 (mean, 30–40) pyloric caecae, and 1–20 basibranchial teeth (Behnke and Zarn 1976).

An excellent map of the historical distribution can be found in Behnke and Benson (1980) and a description of this distribution is in Behnke (1979). Because of habitat loss and introduction of nonnative trouts from the late 1800s to the present, this subspecies is currently rare, and pure populations are virtually nonexistent. The current study began in 1978, in northwestern Colorado, when the Colorado Division of Wildlife recognized a need to determine the purity, distribution, population stability, and abundance of Colorado River cutthroat trout, which was then listed as a threatened subspecies (Miller 1972).

Methods

From 1978 to 1987, 135 streams and 25 lakes were surveyed for the presence of Colorado River cutthroat trout. Selection of streams and lakes to be surveyed was based on computerized inventory data listing waters known to contain cutthroat trout within the historic range of Colorado River cutthroat trout. Records of rainbow trout *Salmo gairdneri* stocking at any time in the history of the water in question automatically eliminated it from consideration because of nearly certain hybridization between the two species. Waters known to have been stocked with cutthroat trout, including nonnative cutthroat trout, were surveyed. Previous inventories revealed few pure populations from lakes and, therefore, survey efforts were concentrated on streams.

Coffelt BP-2 backpack electroshockers and gill nets were used to sample streams and lakes, respectively. An effort was made to sample isolated stream headwaters, because these were considered the most likely locations of pure Colorado River cutthroat trout populations (Behnke and

TABLE 1.—Purity rating criteria used to determine the degree of purity of Colorado River cutthroat trout. Ranges given for scales and pyloric caeca represent mean values. An A represents the most pure population; an F the least pure. Numbers in parentheses indicate numerical value of the letter grade.

Meristic character and spotting pattern	Grade				
	A (1)	B (2)	C (3)	D (4)	F (5)
Scales[a]	180+	168–179	155–167	142–154	120–141
Pyloric caeca	≤40.9	41.0–44.5	44.6–48.5	48.6–53.0	53.1+
Basibranchial teeth[b]	0–10%	10–20%	20–40%	40–75%	75–100%
Spotting variability[c]	Uniform, no variability	Slight variability	Some variability yet still *pleuriticusus*	Quite variable yet still *pleuriticus*	Obvious hybrid spotting

[a] Number of scales two rows above lateral line.
[b] Percent of specimens lacking basibranchial teeth.
[c] Variability in size, number, shape, and position of spots among specimens from the same population.

Zarn 1976). Whenever possible, sampling was performed above barrier falls that precluded upstream migration of nonnative fishes. Barrier falls were considered adequate if they were between 1.5 and 3.1 m high (R.J. Behnke, Colorado State University, personal communication).

Cutthroat trout specimens were preserved in 10% formalin buffered with 0.4% borax (Wernsman 1973; Behnke 1979). A small ventral incision, anterior to the vent, facilitated abdominal penetration of formalin. Only specimens at least 11.5 cm in total length (TL) were preserved, to ensure that they exceeded the size (10.2 cm) at which basibranchial teeth are fully developed (Behnke, personal communication). At least 10 specimens were examined from each water to determine degree of purity of a population (Behnke and Zarn 1976).

Methods used in this study were similar to those described by Schreck and Behnke (1971). A binocular microscope was used to note presence and number (or absence) of basibranchial teeth. Pyloric caeca counts included each separate tip. Scale definition in the two rows above the lateral line was enhanced by scraping away mucus and epidermis and adjusting a light to reflect off scale edges. Occasionally, malachite green was applied to the scales to facilitate counting. Spotting pattern (position, size, shape, and number of spots) was rated on degree of variation between specimens from a single water. Increased variation indicated an increased level of hybridization with nonnative trouts (Behnke and Zarn 1976). The number of gill rakers and the numbers of scales from the anterior margin of the dorsal fin to the lateral line were not counted in the last 5 years of this study because these characters are similar for Colorado River cutthroat trout and rainbow trout.

Purity grades established by Binns (1977) were followed. Letters A (most pure) through F (least pure) designate various degrees of hybridization. Grade E indicates a population not examined by a taxonomist. Grades were assigned to ranges of meristic character means and spotting patterns within a population (Table 1). A numerical value, from one to five, was assigned to each grade. Sums of the numerical value of each meristic character and spotting pattern grade were used to determine final purity rating (4–5 = A, 6–7 = B, 8–10 = C, 11–13 = D, ≥13 = F). Pure populations have sums of four whereas hybridized populations have sums of 13 or more. Pluses and minuses were assigned to the final purity rating to denote where in this range the sum fell, and also, where each meristic mean fell in the purity rating criteria range (Table 1). For example, specimens from Fryingpan Lakes 2 and 3 average 195.1 lateral-series scales, 35.4 pyloric caeca, and 14% of specimens without teeth, and they have an essentially uniform spotting pattern with slight variation. By the purity rating criteria, scales = A = 1, pyloric caeca = A = 1, basibranchial teeth = B = 2, and spotting pattern = A = 1. The sum of these criteria is 5. According to the final purity rating above, 5 = A, but it is a lower-level A; therefore, given the slight variation in spotting pattern, the final rating is A− for the Fryingpan Lakes population.

Results

Among 160 waters surveyed, 21 populations rated A, 38 rated B, 24 rated C, 3 rated D, and 10 rated E were found (Tables 2, 3, and 4). Brook trout *Salvelinus fontinalis* represented the only trout found in 31 waters, and 19 waters were barren of fish. Both Colorado River cutthroat trout and rainbow trout were present in 11 waters, and three waters had only nonnative trouts (rainbow trout, brook trout, and brown trout *Salmo trutta*).

TABLE 2.—Populations of Colorado River cutthroat trout in northwest Colorado with a purity rating of A (most pure). Site numbers correspond to those in Figure 1. Major drainages: C = Colorado River; W = White River. Reasons for ratings downgraded below A+: S = scales; PC = pyloric caeca; BT = basibranchial teeth; SP = spotting pattern.

Site	Location (drainage)	Year sampled	Rating (down-grade)
1	Abrams Creek (C)	1980	A− (PC)
2	Bobtail Creek (C)	1985	A− (PC)
3	Clinton Reservoir (C)	1985	A− (SP)
4	Corral Creek (C)	1984	A− (SP)
5	East Meadow Creek (C)	1982	A+
6	French Gulch (C)	1985	A− (SP)
7	Fryingpan Lakes 2 and 3 (C)	1984	A− (BT, SP)
8	Hack Lake (C)	1980	A (SP)
9	Hahn Creek (W)	1982	A− (SP)
10	Hat Creek (C)	1982	A+
11	Little Vasquez Creek (C)	1983	A+
12	Lost Trail Creek (C)	1982	A− (BT)
13	Mitchell Creek (C)	1984	A+
14	Muddy Creek tributary (C)	1985	A− (SP)
15	Nickelson Creek (C)	1982	A− (SP)
16	Nolan Creek (C)	1982	A (SP)
17	North Fork Swan River (C)	1985	A− (SP)
18	Rocky Fork Creek (C)	1986	A (SP)
19	Snell Creek (W)	1986	A− (SP)
20	Steelman Creek (C)	1985	A− (PC, SP)
21	West Cross Creek (C)	1982	A+

Ninety populations were analyzed for purity, including four from waters that contained rainbow trout. Hybridization between Colorado River cutthroat trout and Yellowstone cutthroat trout *Salmo clarki bouvieri* was evident in 24 of these 90 populations. Hybridization with rainbow trout was observed in another 45 populations. Thirty-seven of the 90 populations were sampled two to four times. Fourteen of these maintained previous ratings, 12 had rating declines, and 7 gained in rating. Eleven of the 12 populations with decreased ratings were previously rated A. One A+ population was not found in subsequent sampling, probably due to overharvest because the entire stream length is easily accessible to anglers. An A− population was replaced by brook trout. Two populations consisting of rainbow trout and Colorado River cutthroat trout were also replaced by brook trout.

Among populations in 18 waters previously stocked with "black-spotted trout" (a collective term used by late 19th- and early 20th-century fish culturists for all presently recognized forms of cutthroat trout: Behnke 1979), including Trappers Lake and Little Trappers Lake, 5 were rated A, 10 B, 2 C, and 1 D. Seven of these 18 populations showed Yellowstone cutthroat trout influence via increased number of basibranchial teeth.

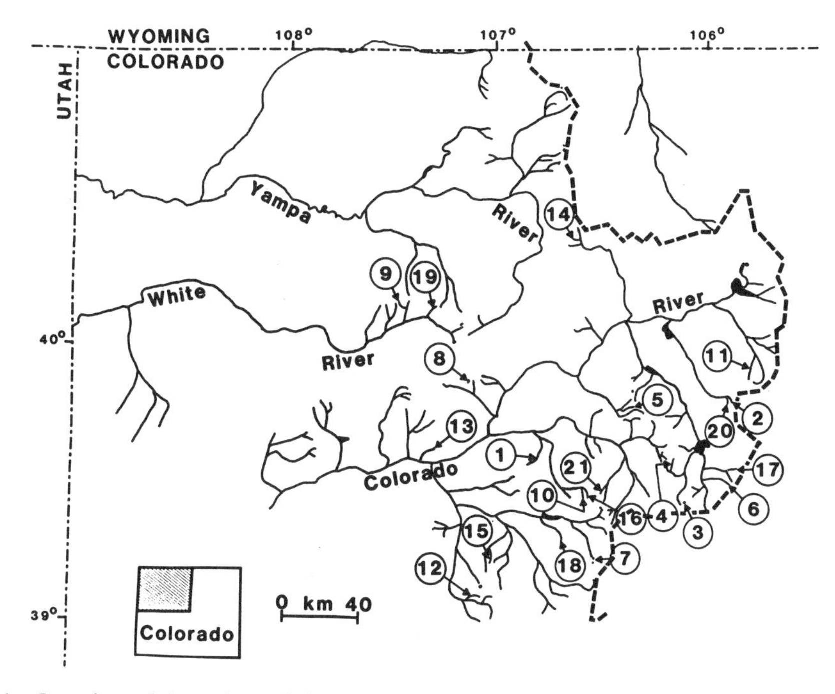

FIGURE 1.—Locations of A-rated populations of Colorado River cutthroat trout. Bold dashed line indicates the Continental Divide. Numbers correspond to sites listed in Table 2.

TABLE 3.—Populations of Colorado River cutthroat trout in northwest Colorado with a purity rating of B (slight indication of hybridization). Site numbers correspond to those in Figure 2. Major drainages: C = Colorado River; N = North Platte River; W = White River; Y = Yampa River. Reasons for ratings downgraded below A+: S = scales; PC = pyloric caeca; BT = basibranchial teeth; SP = spotting pattern.

Site	Location (drainage)	Year sampled	Rating (down-grade)	Site	Location (drainage)	Year sampled	Rating (down-grade)
1	Arapahoe Creek (C)	1981	B (SP)	20	JQS Gulch (C)	1983	B+ (BT, SP)
2	Avalanche Lake (C)	1981	B (BT, SP)	21	Lake 10794 (C)	1981	B+ (S)
3	Berry Creek (C)	1983	B (PC, BT, SP)	22	Lake Diana (Y)	1980	B (PC,BT)
4	Big Beaver Creek (W)	1986	B (PC, SP)	23	Lake of the Crags (Y)	1980	B− (BT)
5	Butler Creek (C)	1982	B+ (BT)	24	Little Skinny Fish Lake (W)	1980	B (BT)
6	Cataract Creek (C)	1981	B+ (SP)	25	Luna Lake (Y)	1980	B+ (SP)
7	Cattle Creek (C)	1985	B (S, PC, SP)	26	Mandall Creek (Y)	1980	B+ (S, SP)
8	Columbine Creek (C)	1981	B (PC, SP)	27	Meadow Creek (C)	1985	B+ (SP)
9	Corral Creek (C)	1983	B+ (S)	28	Miller Creek (C)	1984	B− (BT, SP)
10	Cross Creek (C)	1985	B+ (PC, SP)	29	North Fork Elliott Creek (C)	1981	B (SP)
11	Cunningham Creek (C)	1984	B+ (BT)	30	North Thompson Creek (C)	1981	B (BT, SP)
12	Difficult Creek (C)	1981	B (SP)	31	Oliver Creek (Y)	1984	B− (BT, SP)
13	East Fork Parachute Creek (C)	1983	B+ (BT, SP)	32	Pitkin Creek (C)	1980	B (SP)
14	East Fork Red Dirt Creek (C)	1980	B− (BT)	33	Polk Creek (C)	1986	B+ (SP)
15	East Lake Creek (C)	1984	B− (BT, SP)	34	Porcupine Lake (Y)	1980	B (S,BT)
16	Express Creek (C)	1981	B+ (SP)	35	South Fork Ranch Creek (C)	1984	B− (PC, SP)
17	First Creek (Y)	1981	B− (BT)	36	Spruce Creek (C)	1981	B (BT)
18	Indian Creek (C)	1980	B+ (BT)	37	Ute Creek (C)	1984	B+ (BT, SP)
19	Jack Creek (N)	1978	B (S)	38	West Fork Red Dirt Creek (C)	1980	B− (PC,BT)

Distributions of A, B, and C populations are primarily restricted to headwater streams (Figures 1, 2, and 3). Populations rated A were not found in the Yampa River drainage, and only two were found in the White River drainage (Figure 1). A majority of A, B, and C populations were in the Colorado River drainage.

In 1984, six A populations were sampled for abundance and available habitat (Table 5). Two populations had 12 fish longer than 15 cm TL per

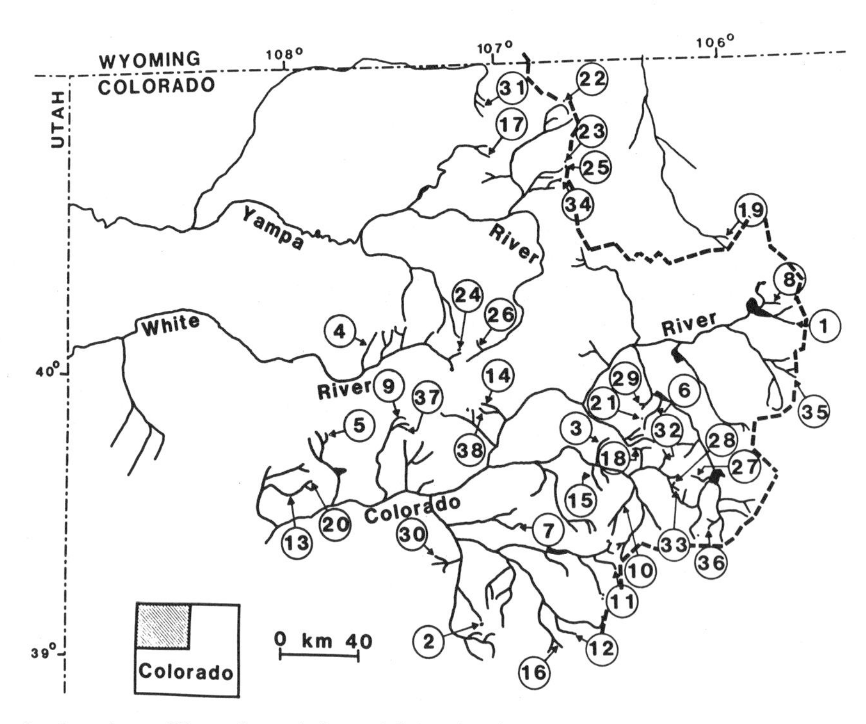

FIGURE 2.—Locations of B-rated populations of Colorado River cutthroat trout. Bold dashed line indicates the Continental Divide. Numbers correspond to sites listed in Table 3.

TABLE 4.—Populations of Colorado River cutthroat trout in northwest Colorado with a purity rating of C (some hybridization). Site numbers correspond to those in Figure 3. Major drainages: C = Colorado River; W = White River; Y = Yampa River. Reasons for ratings downgraded below A+: S = scales; PC = pyloric caeca; BT = basibranchial teeth; SP = spotting pattern.

Site	Location (drainage)	Year sampled	Rating (down-grade)
1	Big Park Creek (C)	1981	C (PC, BT)
2	Black Gore Creek (C)	1982	C (S, BT)
3	Cabin Creek (C)	1985	C (PC, BT, SP)
4	Carter Lake (C)	1980	C (BT)
5	East Middle Fork Parachute Creek (C)	1981	C (S, BT)
6	Hunter Creek (C)	1984	C (BT, SP)
7	Johnson Creek (Y)	1984	C+ (BT, SP)
8	Lake Creek (W)	1978	C (S, BT)
9	Little Green Creek (C)	1985	C− (BT, SP)
10	Lost Creek (W)	1982	C (BT, SP)
11	Lost Dog Creek (Y)	1980	C (S, BT)
12	McCoy Creek (C)	1978	C (BT, SP)
13	Middle Thompson Creek (C)	1981	C+ (BT, SP)
14	Miller Creek (Y)	1984	C− (BT, SP)
15	Northwater Creek (C)	1983	C (S, BT, SP)
16	Poose Creek (Y)	1983	C+ (BT)
17	Possum Creek (C)	1980	C (PC, SP)
18	Red Dirt Creek (C)	1986	C (BT, SP)
19	Roaring Fork Creek (C)	1981	C (BT, SP)
20	Sopris Creek (C)	1985	C− (S, BT, SP)
21	Trail Creek (C)	1984	C+ (BT, SP)
22	Trapper Creek (C)	1983	C+ (BT, SP)
23	Trappers Lake (W)	1987	C (S, BT, SP)
24	Yule Creek (C)	1981	C+ (PC, BT)

100 m of stream, three populations had only 3 fish of such size per 100 m, and one population had less than 1 large fish per 100 m. Suitable habitat available to these six A populations totaled 24 km, a third of which was in Cross Creek. In 1985, however, the Cross Creek population decreased in rating from A+ to B+ due to an increase in both variability of spotting pattern and mean number of pyloric caeca (Table 3). Five B populations and one C population were similarly examined. Cunningham Creek and JQS Gulch, both rated B, had high densities of Colorado River cutthroat trout but the least available habitat among the streams examined.

Discussion

The Colorado River cutthroat trout has never been included in an official federal list of threatened and endangered species, but the Colorado Wildlife Commission listed it as "threatened" in 1976. In 1985, the state delisted the subspecies and categorized it with the fauna of "special concern" (species requiring attention to prevent their decline in status to threatened or endangered) when it was believed that at least 20 stable, relatively pure, populations existed. However, there is reason to be concerned with the current

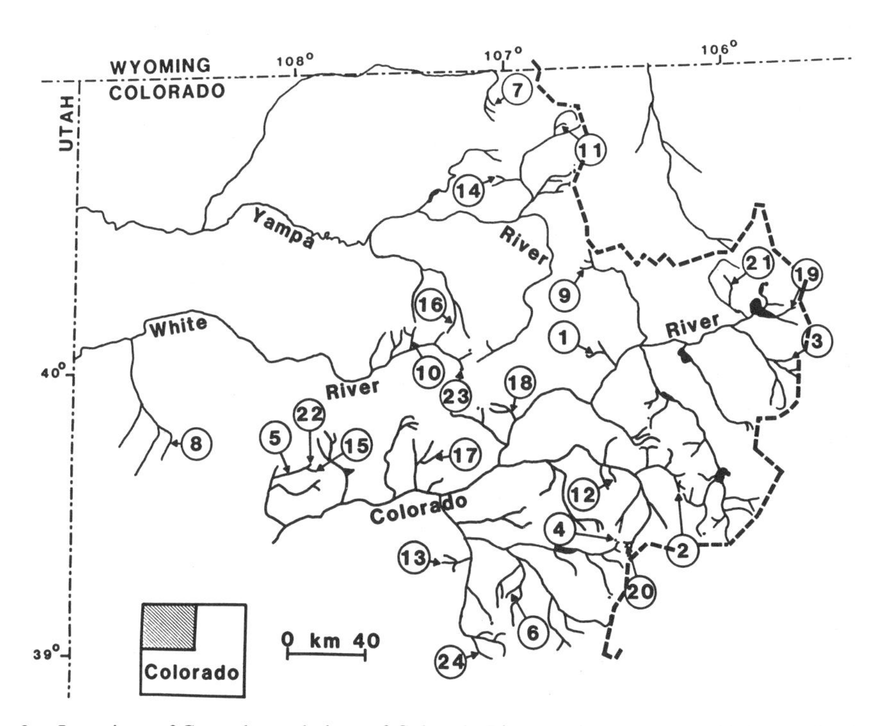

FIGURE 3.—Locations of C-rated populations of Colorado River cutthroat trout. Bold dashed line indicates the Continental Divide. Numbers correspond to sites listed in Table 4.

TABLE 5.—Estimated densities of Colorado River cutthroat trout and extents of their habitat in 12 Colorado streams, 1984. Rating denotes purity: A, most pure; B, slight indication of hybridization; C, some hybridization; TL is total length.

Stream	Rating	Number of fish/100 m		Habitat kilometers
		TL ≥ 10 cm	TL ≥ 15 cm	
Cattle Creek	B	6	3	8.0
Cross Creek	A+	21	12	8.0
Cunningham Creek	B+	72	36	0.8
East Meadow Creek	A+	30	12	4.8
Hat Creek	A+	3	3	1.6
JQS Gulch	B+	60	60	0.8
Miller Creek	B−	18	6	1.6
Mitchell Creek	A+	6	3	3.2
Nolan Creek	A	0	0.3	3.2
Northwater Creek	C	42	42	3.2
South Fork Ranch Creek	B−	24	12	2.4
West Cross Creek	A+	6	3	3.2

stability of the 21 A populations of Colorado River cutthroat trout left in northwestern Colorado. Nine of these have not been sampled in the last 5 years. As stated previously, 11 of 12 resurveyed populations with decreased ratings were previously A populations, and it is possible that the purity status has decreased in some of the nine other populations. Also, the abundance of Colorado River cutthroat trout and available habitat has been substantially reduced. Activities to reestablish this subspecies more broadly in its native range will be nearly impossible if we depend only on remaining A populations located in small headwater streams. Binns (1977) recommended the use of both A and B populations for reintroduction because more relatively pure fish would be available. However, Behnke (1979) recommended using populations with the least hybrid influence (those that still resemble Colorado River cutthroat trout in outward appearance) for restoration programs because there are no populations of this subspecies that have not been exposed to hybridization.

Purity Rating

Variation in spotting pattern is the most subjective criterion of the purity ratings. It is not easily or consistently counted or measured. However, with experience and practice, differences in spot distribution, size, shape, and abundance among fish of a given population become apparent. Behnke and Zarn (1976) described variation of spotting pattern between Colorado River cutthroat trout of the Green River and Colorado River (Little Snake River) basins, and noted all degrees of intermediacy. Genetically pure or essentially pure populations will share a high degree of spotting uniformity, from specimen to specimen, within any geographical area.

Variation in spotting pattern and increased number (Yellowstone cutthroat trout influence) or absence (rainbow trout influence) of basibranchial teeth were most often the first criteria to indicate hybridization (Tables 2, 3, and 4). These results do not support the findings of Behnke and Zarn (1976) or Behnke (1979). Behnke and Zarn (1976) reported that spotting pattern and coloration remained typical of Colorado River cutthroat trout when meristic characters indicated hybridization. Behnke (1979) observed that changes in spotting pattern and coloration were not apparent until 50% of a population lacked basibranchial teeth.

Hybridized Populations

Trappers Lake, the second largest natural lake in Colorado, is near the headwaters of the White River. This lake is the sole source of endemic cutthroat trout eggs used for routine stocking by the Colorado Division of Wildlife in northwestern Colorado. The lake was stocked with Yellowstone cutthroat trout from 1943 through 1950 at rates of 50,000–100,000 fingerlings (50–75 mm TL) each year. Black-spotted trout (any combination of cutthroat trout subspecies or hybrids) were also stocked in 1952 (27,456 fish) and 1965 (20,000 fish). Little Trappers Lake, which drains into Trappers Lake, was stocked with black spotted trout in 10 years between 1954 and 1970.

Rainbow trout were passively introduced into Trappers Lake from a stocking of 500 fish into Little Trappers Lake in 1971. Numerous stockings of catchable rainbow trout in the outlet stream, North Fork White River, also provided a means for entry into Trappers Lake prior to construction of a barrier in 1960. One mature rainbow was captured in the spawning trap in 1958 and four in 1959 (Snyder and Tanner 1960).

Brook trout are currently present in Trappers Lake. An albino brook trout supposedly stocked in Crescent Lake (Colorado River drainage) was also found in Trappers Lake in 1958. This was the result either of an illicit transfer or a misdirected stocking.

Behnke and Benson (1980) noted the lack of hybridization of Trappers Lake Colorado River cutthroat trout with Yellowstone cutthroat trout and rainbow trout but stated that *pleuriticus* could no longer be considered pure because these nonnative trouts were present. Wernsman (1973) also detected no hybridization in a 1971 sample. Pyloric caeca counts were similar to those of Yellowstone cutthroat trout, but Wernsman attributed this to natural variability of Trappers Lake Colorado River cutthroat trout variability and not to hybridization. He apparently counted the number of basibranchial teeth but only recorded their presence or absence, which would give no indication of hybridization with Yellowstone cutthroat trout. Lateral-series scale counts were typical of Colorado River cutthroat trout, averaging 191.

The Trappers Lake population received a purity rating of C following examinations of 18 specimens in 1987 (pyloric caeca: range 29–63, mean 40.5; lateral-series scales; range 147–199, mean 175.5; basibranchial teeth: range 1–42, mean 10.6; highly variable spotting patterns). The upper range of pyloric caeca counts indicates hybridization with rainbow trout (range, 40–70; mean, 55) and the mean suggests possible hybridization with Yellowstone cutthroat trout (range, 31–51; mean, 41.2). Lower scale counts indicate hybridization with rainbow trout. The presence of basibranchial teeth indicate a lack of hybridization with rainbow trout; however, the upper range, 42 teeth, suggests hybridization with Yellowstone cutthroat trout. Highly variable spotting patterns also indicate hybridization. Legendre et al. (1972) noted that exposure of inland cutthroat trout to nonnative rainbow trout typically resulted in hybrid swarms. This appears to be the case in Trappers Lake.

Wernsman (1973) reported finding only three populations of relatively pure Colorado River cutthroat trout in tributaries of the main Colorado River: Cunningham Creek, a tributary to the Fryingpan River; Northwater Creek, a tributary to Parachute Creek; and the headwater source of the Colorado River, Rocky Mountain National Park (RMNP). The latter population was not isolated from nonnative trouts and is now extinct for all practical purposes (Behnke and Zarn 1976). Northwater Creek has been stocked with black-spotted trout and at least one introduction of 1,500 rainbow trout (1976). Behnke (1979) reported cutthroat trout from this creek to be typical Colorado River cutthroat trout with no evidence of hybridization. Twelve specimens taken from Northwater Creek in 1983 exhibited decreased lateral-series scales, highly variable spotting patterns, and absence of basibranchial teeth in 25% of the specimens. This population fell from an A to a C rating (Table 4). Similarly, the Cunningham Creek population fell from A+ in 1973 to B+ in 1984 because 30% of the specimens lacked basibranchial teeth (Table 3).

Potential Brood Stock

Clinton Reservoir contained a relatively pure population of Colorado River cutthroat trout prior to a stocking of Snake River cutthroat trout fingerlings (an undescribed subspecies of *Salmo clarki* indigenous to the upper Snake River, Wyoming) in 1976 and 1977. Because of possible hybridization, fertilized eggs were collected in July 1979 and reared at Glenwood Springs Colorado State Fish Hatchery. These fish (2,000 fry, 2.8 cm TL) were stocked into Timber Lake and Timber Creek (RMNP) in September 1979. In 1980, it was estimated that a ratio of seven Colorado River cutthroat trout to three Snake River cutthroat trout existed in Clinton Reservoir.

Snake River cutthroat trout and greenback cutthroat trout *Salmo clarki stomias,* a subspecies closely related to Colorado River cutthroat trout, were reported to be "interactively segregated" in North Michigan Lake, Colorado by Trojnar and Behnke (1974). These populations live in sympatry and avoid direct competition by modification of behavior patterns from those of allopatric populations. Similar coexistence without hybridization has been documented among Snake River cutthroat trout and millions of rainbow trout and Yellowstone cutthroat trout introduced into the upper Snake River, Wyoming (Trojnar and Behnke 1974).

Though they are not reproductively isolated, it is possible that Snake River cutthroat trout and Colorado River cutthroat trout are similarly segregated in Clinton Reservoir. A purity analysis of Clinton Reservoir Colorado River cutthroat trout conducted in 1985 showed little evidence of hybridization. The Clinton Reservoir population had a purity rating of A− (pyloric caeca: mean, 37.3; lateral series scales: mean, 192.8; 7% of specimens lacked basibranchial teeth). Some spotting-pattern variability reduced the purity rating from A+ to A−. If these two subspecies are segregated, Clinton Reservoir may be a better source of hatchery spawn than Trappers Lake.

Williamson Lakes, California, were stocked with 30,000 fertilized Trappers Lake Colorado River cutthroat trout eggs in 1931. Examination of 21 specimens from Williamson lakes 1, 2, and 3 in 1974 revealed an average of 38.8 pyloric caeca and 188.9 lateral-series scales. Specimens exhibited a range of 2–22 basibranchial teeth (mean, 12.5), and spotting pattern and coloration were typical of this subspecies (Gold et al. 1978). Further examination of 19 specimens collected from Williamson Lake 3 in 1987 showed an average of 38.6 pyloric caeca (range, 31–46), 193 lateral-series scales (range 167–229), and 18.6 basibranchial teeth (range, 2–34). The accepted range of basibranchial teeth for Colorado River cutthroat trout is 1–20. Eight of the 19 specimens had 20 or more teeth, and three had more than 30 teeth. These high basibranchial tooth counts suggest hybridization with Yellowstone cutthroat trout. R. J. Behnke (Colorado State University, personal communication) suggested, however, that a subpopulation of Colorado River cutthroat trout may have formed in Williamson Lake 3. His hypothesis was based on the highly variable nature of Colorado River cutthroat trout, the time that elapsed since the original stocking, the limited number of eggs stocked, and lack of fish passage between Williamson Lake 3 and Williamson lakes 1 and 2. Behnke also suggested that Trappers Lake fish may not have been entirely pure in 1931 when the original eggs were taken. Bench Lake (RMNP) received approximately 300 Colorado River cutthroat trout from Williamson Lakes on August 20, 1987. If a viable brood stock can be developed from these Bench Lake fish, their progeny could be the purest existing Colorado River cutthroat trout available for reintroductions.

Recommendations

Available data indicate a continued decline and tenuous existence of Colorado River cutthroat trout in Colorado. To ensure stability of existing A populations and absence of hybridization, these populations should be reexamined. All remaining A populations should be isolated by fish barriers to preclude contamination with nonnative trouts. Special fishing regulations, fly-and-lure-only and catch-and-release, should be implemented to prevent overharvest of this highly catchable subspecies. Future monitoring of changes in distribution, abundance, purity status, and available habitat is vital for all maintenance and reestablishment endeavors. Illicit stocking may be prevented by public education. Current and future land-use practices must be reevaluated and cooperative agreements with other agencies must be established to prevent continued habitat degradation. Habitat reclamation and removal of all nonnative fishes in sites of historic distribution are necessary for future reintroduction activities. The degree of purity of potential brood-stock populations for reintroductions must be determined. This may include only A-rated populations or all populations with outward appearances of Colorado River cutthroat trout.

Data from the current study indicate that a new source of relatively pure Colorado River cutthroat trout brood stock must be found. The current and only source in Trappers Lake has hybridized with nonnative trouts. Continued stocking of the progeny of these hybridized fish may contaminate any pure populations that may exist.

Acknowledgments

Funding for this study was provided in part by the U.S. Fish and Wildlife Service through the Endangered Species Conservation Act, projects SE-3 and SE-5. Special thanks are extended to Colorado Division of Wildlife personnel, Clee Sealing, Dave Langlois, Mike Grode, Tom Lytle, Gerald Bennett, John Torres, and Robin Knox, and to all the field personnel, for participating in specimen collection and identification of Colorado River cutthroat trout populations. Thanks also are extended to Robert Behnke, Colorado State University, for his direction and guidance throughout this study. I also thank Patrick Martinez and Mary McAfee for reviewing the manuscript.

References

Behnke, R. J. 1979. The native trouts of the genus *Salmo* of western North America. Report to U.S. Fish and Wildlife Service, Denver, Colorado.

Behnke, R. J., and D. E. Benson. 1980. Endangered and threatened fishes of the upper Colorado River basin. Colorado State University, Cooperative Extension Service, Bulletin 503A, Fort Collins.

Behnke, R. J., and M. Zarn. 1976. Biology and management of threatened and endangered western trout. U.S. Forest Service General Technical Report RM-28.

Binns, N. A. 1977. Present status of indigenous populations of cutthroat trout, *Salmo clarki,* in southwest Wyoming. Wyoming Game and Fish Department, Fishery Technical Bulletin 2, Cheyenne.

Cope, E. D. 1872. Report on the recent reptiles and fishes of the survey, collected by Campbell Carrington and C. M. Dawes. Pages 467–476 *in* F. V. Hayden, editor. Fifth annual report, United States Geological Survey of Montana and portions of adjacent territories, 5th annual report. U.S. Government Printing Office, Washington, D.C.

Gold, J. R., G. A. E. Gall, and S. J. Nicola. 1978. Taxonomy of the Colorado cutthroat trout *(Salmo clarki pleuriticus)* of the Williamson lakes, California. California Fish and Game 64:98–103.

Johnson, J. E. 1976. Status of endangered and threatened fish species in Colorado. U.S. Bureau of Land

Management, Technical Note T/N 280, Denver, Colorado.

Jordan, D. S. 1891. Report of explorations in Colorado and Utah during the summer of 1889, with an account of the fishes found in each of the river basins examined. U.S. Fish Commission Bulletin 9:1–40.

Legendre, P., C. B. Schreck, and R. J. Behnke. 1972. Taximetric analysis of selected groups of western North American *Salmo* with respect to phylogenetic divergences. Systematic Zoology 21:292–307.

Miller, R. R. 1972. Threatened freshwater fishes of the United States. Transactions of the American Fishery Society 101:239–252.

Schreck, C. B., and R. J. Behnke. 1971. Trouts in the upper Kern River basin, California, with references to systematics and evolution of western North American *Salmo*. Journal of the Fisheries Research Board of Canada 28:987–998.

Snyder, G. R., and H. A. Tanner. 1960. Cutthroat trout reproduction in the inlets to Trappers Lake. Colorado Department of Game and Fish, Technical Bulletin 7, Fort Collins.

Trojnar J. R., and R. J. Behnke. 1974. Management implications of ecological segregation between two populations of cutthroat trout in a small Colorado lake. Transactions of the American Fisheries Society 103:423–430.

Wernsman, G. 1973. The native trout of Colorado. Master's thesis. Colorado State University, Fort Collins.

American Fisheries Society Symposium 4:90–92, 1988

Rio Grande Cutthroat Trout Management in New Mexico

JEROME A. STEFFERUD

U.S. Forest Service, Santa Fe National Forest, Post Office Box 1689
Santa Fe, New Mexico 87504, USA

Abstract.—Populations of Rio Grande cutthroat trout *Salmo clarki virginalis* have been identified in 39 streams in New Mexico. The New Mexico Department of Game and Fish has set goals for management of this subspecies that include protection of existing populations through restrictions on introductions of nonnative salmonids, investigations to locate new populations and to determine their genetic purity through electrophoretic techniques, and identification of restoration sites. Personnel from Carson and Santa Fe national forests have identified habitat improvement of streams containing Rio Grande cutthroat trout as a priority wildlife activity. The state and federal agencies have agreed to construct migration barriers to protect existing populations against invasion by nonnative trouts.

The Rio Grande cutthroat trout *Salmo clarki virginalis* is endemic to the upper Rio Grande basin in Colorado and northern New Mexico and the Pecos River in New Mexico (Behnke 1988, this volume). It has also been found in a few headwater streams of the Canadian River in Colorado and New Mexico, where its native status is still questionable (Behnke 1979). Populations within each of the three major drainages can be distinguished from each other through meristic (Behnke 1980, 1982) and electrophoretic analyses (M. Hatch, New Mexico Department of Game and Fish, personal communication).

Like other inland cutthroat trout subspecies, the Rio Grande cutthroat trout has been extirpated from a large portion of its native range during the past century (Behnke 1979). Introductions of nonnative trouts that either interbreed and hybridize with the native species or directly outcompete or replace them have severely restricted the distribution and abundance of Rio Grande cutthroat trout. Additionally, habitat degradation, caused by conflicting land uses and stream dewatering, has contributed to the present limited distribution; genetically pure populations are found only in small, isolated headwater streams (Propst and McInnis 1975).

The Rio Grande cutthroat trout was 1 of 10 subspecies of cutthroat trout placed on the U.S. Government Federal list of threatened species (Anonymous 1973), but it was not listed under the Endangered Species Act of 1973. The Endangered Species Committee of the American Fisheries Society considers it a species of "special concern" because of the destruction of its habitat and its hybridization with and competition from other species (Deacon et al. 1979; Johnson 1987). In New Mexico, the Rio Grande cutthroat trout is considered a game species subject to fishing regulations established by the State Game Commission.

Numerous investigations during the past decade by staff and biologists from the New Mexico Department of Game and Fish and from the Carson and Santa Fe national forests have revealed that Rio Grande cutthroat trout populations, although still uncommon throughout their native range, are more abundant than previously thought (New Mexico Department of Game and Fish 1987). Currently, populations of Rio Grande cutthroat trout have been identified in 39 streams in New Mexico; however, fewer than one-third of those streams are protected from invasion by nonnative trout by migration barriers.

Management

In part, the basic mission of the New Mexico Department of Game and Fish is to "Preserve the natural diversity and distribution patterns of the State's native ichthyofauna" (New Mexico Department of Game and Fish 1987). New Mexico has established a management program for maintaining and enhancing existing Rio Grande cutthroat trout populations and for establishing new populations within the fish's native range. The objective of the program is to ensure that Rio Grande cutthroat trout remain an integral part of the state's fisheries and do not diminish to the point of requiring special protective regulations. The state's management program has been incorporated into land and resource management plans of the Santa Fe (U. S. Forest Service 1987) and Carson national forests (U. S. Forest Service 1986). In general, the management programs consist of (1) protection of existing populations; (2)

restoration into suitable waters; (3) inventory, monitoring, and genetic analysis; (4) artificial propagation and brood-stock maintenance and (5) habitat protection and enhancement.

Protection of Existing Populations

Nonnative salmonids will not be introduced into those portions of streams occupied by Rio Grande cutthroat trout, and barriers will be constructed to exclude nonnative salmonids from Rio Grande cutthroat trout habitats. Efforts will be made to locate natural barriers or sites that can be modified to serve as barriers. Nonnative trout will be removed from streams that have barriers if such removal is feasible.

Of the 39 streams that contain Rio Grande cutthroat trout, 27 are in the Rio Grande basin and 6 each are in the Pecos and Canadian basins. Extant stocks will not be mixed between basins. The Canadian basin populations are in the most precarious situation; four of the six streams are on private land, and one of the two that are in Carson National Forest has been invaded by brown trout *Salmo trutta*. The other national forest population, in McCrystal Creek, appears secure under present management, and the habitat is in good condition (Carson National Forest, unpublished data). Carson National Forest has restricted public vehicular access to a point 5 km from the creek and a portion of the stream is being fenced to protect it from grazing by cattle and elk.

All six streams in the Pecos drainage are in Santa Fe National Forest, and three have natural barriers that prevent upstream movement of nonnative trouts. In the other three streams, the absence of nonnative trouts is probably the result of low water temperatures, stream turbulence, and other undetermined parameters (Propst and McInnis 1975). All streams are in wilderness or other remote areas and in terrain unsuitable for cattle grazing.

Rio Grande cutthroat trout populations have been found throughout the upper Rio Grande basin in the Sangre de Cristo, Jemez and Brokeoff mountain ranges. During the past decade, six barriers have been constructed to protect known populations of Rio Grande cutthroat trout, and one stream system was chemically treated to remove nonnative trout.

Restoration into Suitable Waters

The suitability of streams as sites for restoration of Rio Grande cutthroat trout populations will be assessed; however, this program is likely to have a low priority until many of the existing populations are protected. Adverse public opinion has caused chemical renovation projects to be delayed or cancelled. Additionally, it is difficult to justify restoration projects when existing populations are unprotected.

Inventory, Monitoring and Genetic Analysis

Unsurveyed streams within the native range of Rio Grande cutthroat trout, particularly the headwater sections, will be inventoried by New Mexico Game and Fish or U.S. Forest Service personnel to determine distribution, relative abundance, genetic purity, and habitat quality. The genetic profile of Rio Grande cutthroat trout in the major basins in New Mexico has been determined (Hatch, personal communication), and new populations will be tested against that profile. Surveys will normally be conducted during the course of other work (e.g., during inventories for timber sales, roads, and wilderness evaluations).

Monitoring of habitat conditions will be done by Forest Service personnel as described in the forest plans for the areas (U. S. Forest Service 1986, 1987). Briefly, water quality will be monitored through application and analysis of "best management practices" designed to minimize erosion caused by forest activities. The objective will be to meet state water quality standards. The macroinvertebrate community will be sampled in selected streams in order to determine long-term trends in water and habitat quality as measured by the biotic condition index and community tolerance quotient (Winget and Mangum 1979).

Direct monitoring of populations in selected streams will be done to determine baseline conditions of community abundance and biomass. Biannual monitoring in Rio Grande cutthroat trout streams in Carson National Forest has been carried out since 1983.

Artificial Propagation and Broodstock Maintenance

New Mexico Department of Game and Fish plans to develop Rio Chamita and its tributaries within their Sargent Ranch property as a source of brood fish for Rio Grande cutthroat trout. The progeny of these fish will be used in reestablishing the subspecies in streams in the Rio Grande drainage. The department also plans to investigate the use of Rio Grande cutthroat trout to provide self-sustaining fisheries in remote areas.

In 1987, Mescalero National Fish Hatchery near Alamogordo, New Mexico, collected 22 Rio Grande cutthroat trout adults out of Indian Creek in the Sacramento Mountains. Fish were spawned and 800–1,000 swim-up fry have been produced. A brood stock will be developed in order to supply cutthroat trout to stock steams on the Mescalero Indian Reservation (J. Burton, U. S. Fish and Wildlife Service, personal communication).

Habitat Protection and Enhancement

Protection and enhancement of stream habitats have been shown to increase trout abundance and biomass in western trout streams (Hall and Baker 1982) if the program is properly designed and incorporates an understanding of the particular fish population, habitat, and cultural setting (White 1986). If the hypotheses presented by Griffith (1988, this volume) are valid (i.e., that cutthroat trout in optimal habitat may be able to withstand competition from introduced salmonids, and that replacement rather than competition is the mechanism causing decline of many interior stocks of cutthroat trout), then habitat protection and enhancement may be extremely powerful tools in the future maintenance of Rio Grande cutthroat trout stocks.

Rio Grande cutthroat trout have been identified as a management indicator species in the forest plans of the Carson and Santa Fe national forests (U. S. Forest Service 1986, 1987). Indicator species are selected to help monitor the effects of planned management activities of the forests. As such, streams containing Rio Grande cutthroat trout will receive emphasized management consideration during project planning and will be monitored during project implementation to ensure compliance with habitat and water quality standards.

Carson National Forest has installed protective fencing or riparian pastures on Tio Grande, McCrystal, and Commanche creeks and has constructed instream structures on Rio Costilla, Tio Grande, and Policarpio creeks. Approximately 550 fish habitat improvement structures are planned to be installed in Rio Grande cutthroat trout streams in Carson National Forest during the next decade.

Acknowledgments

I acknowledge the contributions that Michael D. Hatch, New Mexico Department of Game and Fish, made to this paper. I also thank R. E. Gresswell and three anonymous reviewers for their constructive comments on the manuscript.

References

Anonymous. 1973. Threatened wildlife of the United States. U. S. Bureau of Sport Fisheries and Wildlife Resource Publication 114.

Behnke, R. J. 1979. The native trouts of the genus *Salmo* of western North America. Report to U. S. Fish and Wildlife Service, Denver, Colorado.

Behnke, R. J. 1980. Report on collections of cutthroat trout from north-central New Mexico. Report to New Mexico Department of Game and Fish, Santa Fe.

Behnke, R. J. 1982. Evaluation of 1982 collections of cutthroat trout from New Mexico. Report to New Mexico Department of Game and Fish, Santa Fe.

Behnke, R. J. 1988. Phylogeny and classification of cutthroat trout. American Fisheries Society Symposium 4:1–7.

Deacon, J. E., G. Kobetich, J. D. Williams, and S. Contreras. 1979. Fishes of North America endangered, threatened, or of special concern: 1979. Fisheries (Bethesda) 4(2):29–44.

Griffith, J. S. 1988. Review of competition between cutthroat trout and other salmonids. American Fisheries Society Symposium 4:134–140.

Hall, J. D., and C. O. Baker. 1982. Rehabilitating and enhancing stream habitat: 1. Review and evaluation. U. S. Forest Service General Technical Report PNW-138.

Johnson, J. E. 1987. Protected fishes of the United States and Canada. American Fisheries Society, Bethesda, Maryland.

New Mexico Department of Game and Fish. 1987. Operation plan: aquatic management of New Mexico wildlife, 1987-1995. Santa Fe.

Propst, D. L., and M. A. McInnis. 1975. An analysis of streams containing native Rio Grande cutthroat trout, *Salmo clarki virginalis*, on the Santa Fe National Forest. Western Interstate Commission for Higher Education, Boulder, Colorado.

U. S. Forest Service. 1986. Carson National Forest plan. Taos, New Mexico.

U. S. Forest Service. 1987. Santa Fe National Forest plan. Santa Fe, New Mexico.

White, R. J. 1986. Physical and biological aspects of stream habitat management for fish: the primacy of hiding/security cover. Pages 241–265 *in* J. G. Miller, J. A. Arway, and R. F. Carline, editors. Proceedings of the 5th trout stream habitat improvement workshop. Pennsylvania Fish Commission, Harrisburg.

Winget, R. N., and F. A. Mangum. 1979. Biotic condition index: integrated biological, physical, and chemical stream parameters for management. U. S. Forest Service, Intermountain Region, Ogden, Utah.

American Fisheries Society Symposium 4:93–106, 1988

Status, Life History, and Management of the Lahontan Cutthroat Trout

ERIC R. GERSTUNG

*Inland Fisheries Division, California Department of Fish and Game, 1416 Ninth Street
Sacramento, California 95814, USA*

Abstract.—The Lahontan cutthroat trout *Salmo clarki henshawi* is believed to have occupied at least 6,100 km of stream habitat and 135,000 hectares of lake habitat in its historic range. Some lake populations such as those in Lake Tahoe, California–Nevada and in Pyramid and Walker lakes in Nevada were large enough to support thriving commercial fisheries. Pure self-sustaining stocks of Lahontan cutthroat are now limited to headwater streams within the Humboldt River drainage, to Summit and Independence lakes, and to several small tributaries of the Truckee, Carson, and Walker rivers. A small number of additional populations have been established outside the Lahontan Basin in both California and Nevada by transplants and hatchery production in each of these states. Self-sustaining Lahontan cutthroat trout populations are known to occur in 100 locations, representing about 7% of the historic stream habitat and 0.4% of the historic lake habitat. Competition from and hybridization with introduced nonnative trout, the construction of dams and diversions on important spawning streams, and water quality deterioration are largely responsible for the decline of Lahontan cutthroat trout. Restoration efforts in California and Nevada will involve eliminating nonnative trout from selected streams, restocking with endemic stocks where possible, and improving and protecting habitat. Hatchery-reared stocks will continue to be utilized in various lake and reservoir recreational fishing programs, including those existing in Pyramid and Walker lakes.

Lahontan cutthroat trout *Salmo clarki henshawi* were endemic to the Pleistocene lake, Lake Lahontan, which occupied 13,000 km^2 within present-day northwestern Nevada and northeastern California during its maximum level some 25,000 years ago (McAfee 1966; Moyle 1976). Lake Lahontan fluctuated widely throughout the centuries in response to climatic changes. Following a long drying trend, final desiccation occurred 5,000-9,000 years ago (Benson 1978). Pyramid and Walker lakes are remnants of Lake Lahontan. Four major stream systems, including the Truckee, Carson, Walker, and Humboldt rivers, also remain.

The Lahontan cutthroat trout was extirpated from much of the native range following European settlement. The decline of the Lahontan cutthroat trout and its causes have been described (Juday 1907; Snyder 1917; Sumner 1939; Scott 1957; Wheeler 1974; Behnke 1979; Townley 1980; Coffin 1983) and are summarized in this paper. This paper also reviews the current status of Lahontan cutthroat trout populations in California and Nevada and the management efforts being made to maintain and expand these populations.

Historic Distribution and Decline

When European settlers first arrived in the Lahontan drainage over a century ago, Lahontan cutthroat trout reportedly existed in waters throughout the basin (Dadd 1869; Jordan and Henshaw 1878; Rutter 1902; Snyder 1917; La Rivers 1962; McAfee 1966; Behnke 1979; Coffin 1983). It is suspected that Lahontan cutthroat trout occurred in most of the cooler perennial streams downstream from impassable falls. At least 6,100 km of stream habitat were in this category (Figure 1; Table 1). In addition, Lahontan cutthroat trout are known to have occurred in Tahoe, Cascade, Fallen Leaf, Independence, Donner, Pyramid, Winnemucca, Walker, Upper Twin, Lower Twin, and Summit lakes. These 11 lakes had a combined surface area of 135,000 hectares. The majority of Lahontan cutthroat trout habitat existed in the subbasins of the Truckee, Carson, Walker, and Humboldt rivers, in Honey Lake, and in the drainage encompassing the Smoke Creek and Black Rock deserts.

Truckee River Subbasin

Within the Truckee River system, the largest Lahontan cutthroat trout populations occurred in Lake Tahoe and Pyramid and Winnemucca lakes, where the fish served as a major food source for local Paiute Indians and supported important commercial fisheries (Sumner 1939). During the late 1800s, 45,450–90,000 kg of Lahontan cut-

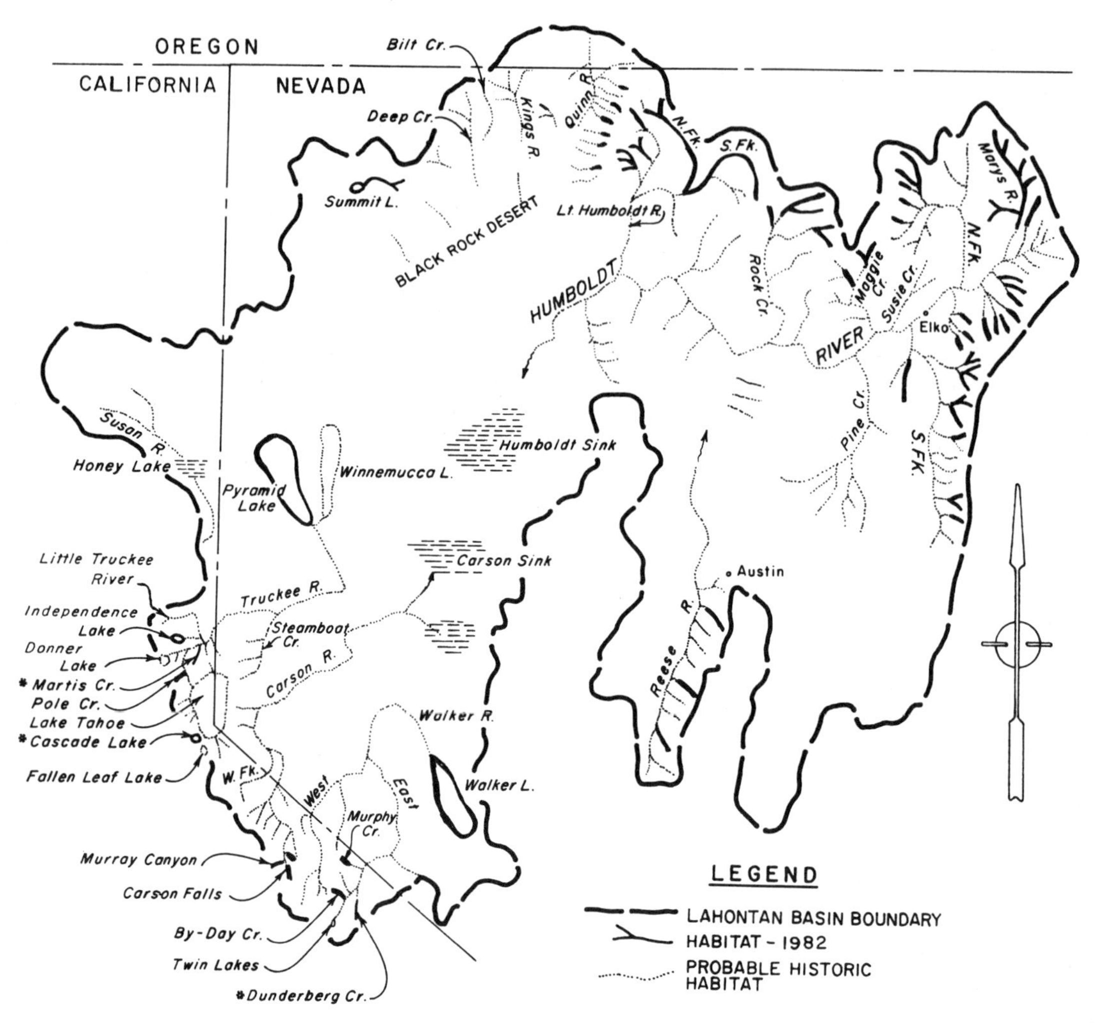

FIGURE 1.—Present and probable historic habitats of the Lahontan cutthroat trout. Asterisks denote introgressed populations.

throat trout were harvested annually in Pyramid Lake and from Truckee River and were transported by wagon and railroad to Nevada towns and mining camps; surplus fish were shipped to San Francisco and other cities. Review of commercial catch records suggests that annual spawning runs up the Truckee River may have consisted of tens of thousands of fish (Behnke 1974). For example, Wells Fargo Express and railroad freight lines shipped over 46,000 kg of fish from towns along Truckee River in 1900 (Bailey 1978).

In addition to the productive commercial fishery, trophy-sized fish in the lake and river attracted anglers from throughout the world (Wheeler 1974). Catches of 4.5–9.0-kg Lahontan cutthroat trout were not uncommon during the 1920s. The world record sport-caught Lahontan cutthroat trout weighed 18.6 kg, and an even larger fish, estimated to weight 27.3 kg, was caught in the commercial fishery (Wheeler 1974).

The Pyramid Lake fishery began to decline in 1906, slowly at first, then more rapidly during the 1920s. The last notable spawning run was reported in 1928, and by 1935 boat angling on the lake was poor (Sumner 1939). Spawning runs up the river ceased after 1938, and by the early 1940s Lahontan cutthroat trout were extinct in Pyramid Lake (Trelease 1952).

Early literature including fish commission reports for Nevada and California, noted that rampant poaching, uncontrolled exploitation, river obstructions, and sawmill and pulp mill pollution

TABLE 1.—Probable historic Lahontan cutthroat trout waters and known waters currently supporting self-sustaining populations within the Lahontan basin.

Water	Historic habitat		Known occupied habitat (1982)		
			Self-sustaining		Artificially maintained lake habitat (hectares)
	Stream (km)	Lake (hectares)	Stream (km)	Lake (hectares)	
Honey Lake drainage	250	None	None	None	None
Truckee system	600	115,000	5	284	45,000
Carson system	500	None	17	None	None
Walker system	600	20,000	6	None	15,000
Humboldt system	3,500	None	444	None	None
Black Rock Desert system (including Quinn River)	640	None	8+	None	None
Summit Lake drainage	10	240	10	240	None
Total	6,100	135,000	490	524	60,000

were having an adverse affect on the Pyramid Lake fishery (Behnke and Zarn 1976). However, it was the construction of the Newlands Irrigation Project in 1905 that ultimately extirpated the Pyramid Lake population of Lahontan cutthroat trout. Derby Dam, operated by the U.S. Bureau of Reclamation on the Truckee River 67 km above Pyramid Lake, diverts most of the flow of the Truckee River into the Carson River basin, except during periods of exceptional runoff. As a result, fish are blocked at Derby Dam, and stream flows below Derby Dam are not adequate for cutthroat trout reproduction (USFWS 1979). Also, the water level in Pyramid Lake dropped 21.3 m between 1906 and 1968, exposing a delta at the mouth of the river that obstructs fish attempting to migrate from the lake (Bailey 1978). Shallower Winnemucca Lake, which shared the natural flow of the Truckee River with Pyramid Lake, dried up totally in 1938 (Sumner 1939).

Lake Tahoe was also famous for its sport and commercial fishery for Lahontan cutthroat trout. Juday (1907) reported that the sport and commercial catches approached 33,000 kg in some years. Accounts from newspapers at the end of the 19th century indicated that daily sport catches of 50–100 Lahontan cutthroat trout per angler were common (Scott 1957).

Each spring (March through July) thousands of adult Lahontan cutthroat trout migrated upstream to spawn in the accessible tributaries to Lake Tahoe (Shebley 1929). Market fishermen established permanent traps on the major tributaries where the entire run could be taken. The more productive tributaries such as Taylor Creek supported spawning runs of up to 7,000 cutthroat trout annually (Hunt 1910). Fish for market were also taken by gill nets and beach seines, some of which were as long as 800 m (Scott 1957).

As a result of overfishing and damage to spawning tributaries caused by pollution, logging, and diversions, the fishery at Lake Tahoe had begun to decline by the 1880s (Shebley 1894; Miller 1951). From 1882 through 1938, the California Fish and Game Commission conducted egg-taking operations at the lake. During this period, 2–6 million eggs were obtained each season from adults trapped in tributary streams (Shebley 1929). Although some of the resulting progeny were returned to the lake, the majority were stocked in other waters, a practice which contributed to the decline in the cutthroat trout fishery. Although commercial fishing was banned from the lake in 1917, the cutthroat trout fishery continued to decline and, in 1938, the last tributary spawning runs were observed (Curtis 1938; Scott 1957; Cordone and Frantz 1966).

Lake trout *Salvelinus namaycush*, as well as rainbow trout *Salmo gairdneri* and brown trout *S. trutta*, all of which were introduced before the turn of the century, had become well established by the 1930s and probably helped seal the fate of the Lake Tahoe population of Lahontan cutthroat trout. Evermann and Bryant (1919) observed that Lahontan cutthroat trout were seldom found in tributaries where rainbow, brown, and brook trout *Salvelinus fontinalis* were established and that the remaining cutthroat trout spawning runs at Lake Tahoe were being maintained by hatchery fish. In nearby Fallen Leaf Lake, where habitat changes and overfishing had not been a major problem, introduced lake trout had largely displaced the native cutthroat trout by 1920 (James 1921).

In an unsuccessful effort to restore the Lake Tahoe fishery, the California Department of Fish and Game (CDFG) planted nearly 1 million fingerling and yearling Lahontan cutthroat trout in Lake

Tahoe between 1956 and 1962. The failure of this and other efforts to reestablish Lahontan cutthroat trout suggests the niche formerly filled by cutthroat trout was now occupied by nonnative trout (Cordone and Frantz 1968).

Nonnative trout were also introduced in other portions of the Truckee River system with the same consequences to Lahontan cutthroat trout. Rainbow and brown trout were present throughout the Truckee River Basin by 1915 and, by 1930, few waters within the drainage still contained cutthroat trout (Snyder 1917; CDFG, unpublished files). In 1960, populations of pure Lahontan cutthroat trout in the Truckee River basin were limited to Pole Creek, a small tributary to the Truckee River, and to Independence Lake (Behnke 1979). The Pole Creek population was entirely displaced by brook trout a decade later. The Independence Lake population declined from an estimated 2,000–3,000 spawners in 1894 (Shebley 1894; Smith 1898; Welch 1929) to less than 100 spawners annually since 1960 (Lea 1968; Gerstung 1986). (The historic estimate includes the combined take of fish harvested in the spawning tributary by market fishermen and of spawners trapped at an egg-taking weir; estimates made since 1960 are based on weekly counts of adult fish on the spawning grounds.)

Lahontan cutthroat trout have been reestablished by transplants into Pole, Grey, and Hill creeks within the Truckee River drainage (M. Warren, Nevada Department of Wildlife, personal communication). Hill and Grey creeks, however, are subject to flash floods; thus, cutthroat trout populations face an uncertain future. Recent transplants of Lahontan cutthroat trout into the headwaters of Bronco, Deep, and Martis creeks failed to establish self-sustaining fisheries. Additional transplants into tributaries of the Truckee River have been proposed (Gerstung 1986).

Carson River Subbasin

Historic distribution of Lahontan cutthroat trout in the Carson River drainage probably included most of the drainage downstream from Carson Falls on the East Fork and downstream from Faith Valley on the West Fork (Figure 1). It is doubtful that endemic lacustrine populations existed during historic times. Lahontan cutthroat trout from the West Fork Carson River were stocked in the nearby Blue Lakes in 1864 (Evermann 1906). The CDFG maintained a brood stock from this source in Heenan Lake until 1980 when the population was phased out and replaced with another broodstock from Independence Lake. The Nevada Department of Wildlife (NDW) continues to maintain Heenan Lake (Carson River) brood stock in Marlette Lake (King 1982).

Self-sustaining populations of genetically pure Lahontan cutthroat trout no longer occupy historic habitat within the Carson River subbasin. Despite some habitat degradation, resulting from agricultural, mining, and logging activities, the introduction of nonnative trout before the turn of the century is probably responsible for the extirpation of Lahontan cutthroat trout within the Carson River drainage (McAfee 1966). Although naturally occurring Lahontan cutthroat trout populations have been eliminated from the Carson River drainage, small populations have become established in the formerly fishless headwaters of the East Fork Carson River and three tributaries above impassable falls. Sheepherders, who once frequented the region, may have planted these streams with endemic stocks of Lahontan cutthroat trout collected below the falls; however, CDFG planting records reveal that these streams were also planted with hatchery-reared fry, most of which were of Blue Lake (Carson River) origin. It is not known whether these stocked fish survived and reproduced. Meristic and electrophoretic analyses indicate that none of the fish collected from these waters are recognizably introgressed with rainbow trout (Behnke 1979; Loudenslager and Gall 1980a, 1980b, 1980c). For management purposes the CDFG is treating the Lahontan cutthroat trout populations occurring in the headwaters of the East Fork Carson River as Carson River stock to be used in restocking reclaimed tributaries of the Carson River (Gerstung 1986).

Walker River Subbasin

Historically, Lahontan cutthroat trout occurred throughout the Walker River drainage from the headwaters in California downstream to Walker Lake, Nevada (Figure 1). Lahontan cutthroat trout also occurred in Upper and Lower Twin lakes on the Robinson Creek tributary (Vestal 1950). Smith and Needham (1935) reported that "great schools" of Lahontan cutthroat trout formerly migrated upstream from Walker Lake during spring spawning migrations and provided excellent fishing. The Fremont party in 1844 observed that Indians maintained numerous fishing weirs along Walker River where so-called "salmon trout" were speared or dipped from the water (Fremont 1845). These Indians developed a market fishery on the lower river during the 1880s that lasted until the early 1930s (Johnson 1974).

Before the turn of the century, agricultural development and associated irrigation diversions began taking their toll of the cutthroat trout. One early newspaper account noted that fish were having difficulty passing over irrigation dams in the lower river valleys, particularly during dry years, and that their offspring often ended up in irrigated fields (Johnson 1974). A decline in the lake fishery was apparent by 1911, and the last notable runs up the river occurred in the late 1920s (Allan 1958).

Completion of Bridgeport Reservoir in 1924 blocked spawning migrations up the East Walker River (Wedertz 1978). Although a fish ladder was present, it was not effective (Vestal 1950). In 1935, completion of Weber Dam on the main stem below the forks ended all migration up the Walker River (La Rivers 1962). The remainder of the river downstream from the dam was unsuitable for cutthroat trout reproduction. Consequently, the cutthroat trout population in Walker Lake was reduced to a few large fish by the 1940s (Allen 1958). During 1949, 39 Lahontan cutthroat trout were seined from the lake near the mouth of the river and transported to Verdi Hatchery to establish a brood stock (Wheeler 1974). Progeny of the brood stock are used for stocking Walker Lake, which still supports a sport fishery (Allan 1958; Sevon 1982, 1987).

As a result of continued water quality deterioration, the Walker Lake fishery faces an uncertain future. Following an intensive limnological study, Koch et al. (1977) concluded that Walker Lake is suffering from accelerated eutrophication, increasing ionic concentrations, and increasing hypolimnetic dissolved oxygen depletions. Since the early 1900s the lake level has dropped 40 m, the volume has decreased by two-thirds, total dissolved solids have more than doubled (to 10,500 mg/L), and alkalinity, (resulting from high bicarbonate levels, now exceeding 2,870 mg/L) has more than doubled (Koch et al. 1977). Irrigation return flows continue to add nutrients to the lake, and diversions remove almost the entire inflow during dry years. The lake level is declining at the rate of about 0.5–1.0 m/year (Koch et al. 1977). During the late summer, particularly during dry climatic cycles, conditions for cutthroat trout survival in Walker Lake are marginal (Koch et al. 1977; Sevon 1987).

Introduction of nonnative trout into Walker River and its tributaries during the late 1800s resulted in the gradual extirpation of Lahontan cutthroat trout from the drainage. However, a number of tributaries, such as Desert and Dunderberg (Dog) creeks, supported pure Lahontan cutthroat trout populations as late as the 1970s (Behnke and Zarn 1976; T. C. Frantz, NDW, personal communication). With the exception of the artificially propagated Walker Lake brood stock, only a single endemic population of Lahontan cutthroat trout now exists in the Walker River system. This population occurs in By-Day Creek, tributary to the East Walker River, which, perhaps because of its small size, was never stocked with trout. Since 1977, Lahontan cutthroat trout from By-Day Creek have become established by transplant in Murphy Creek, another Walker River tributary (D. Wong, CDFG, personal communication). The By-Day Creek population will be used as a brood-stock source for establishing the Walker River subbasin race in other Walker River tributaries (Gerstung 1986).

Humboldt River Subbasin

The Humboldt River drainage supports most of the remaining fluvial populations of endemic Lahontan cutthroat trout within the Lahontan Basin. These fish differ meristically from those in the Truckee, Carson, and Walker river drainages (i.e., their scales are smaller and they have fewer gill rakers); these differences have led some taxonomists to speculate that the populations warrant subspecific recognition (Behnke and Zarn 1976; Behnke 1988, this volume). Behnke (1979) hypothesized that cutthroat trout in the Humboldt River basin became isolated some 8,000 years ago, following the desiccation of Lake Lahontan, and became better adapted to living in a fluvial environment than the lacustrine cutthroat trout in the western Lahontan basin. Loudenslager and Gall (1980a) contended that differences among meristic characters were not sufficient to warrant subspecies separation; however, they conceded that the electrophoretic data suggested that cutthroat trout from the Humboldt River system could be considered a "microgeographic race" of Lahontan cutthroat trout.

Coffin (1981) speculated that Lahontan cutthroat trout within the Humboldt River system may have once existed in as many as 300 coldwater streams comprising about 3,500 km of habitat. Accounts from early residents indicated that Lahontan cutthroat trout occupied the upper third of both the Humboldt River (upstream from Battle

Mountain) and its longest tributary, the Reese River (from the Reese River Canyon upstream), and other suitable tributaries (Dadd 1869; Durrant 1935; Coffin 1981).

Lahontan cutthroat are now found in about 12% of their historic habitat in the Humboldt River drainage and are largely restricted to upper-basin tributaries at elevations above 1,500 m (Figure 1). As of 1987, Lahontan cutthroat trout have been collected from 83 streams, 444 km of combined habitat (P. D. Coffin, NDW, personal communication). Additionally, Lahontan cutthroat are likely to exist in 12 or more streams that they occupied when the streams were last surveyed during the 1950s (Frantz and King 1958; Behnke 1960; Coffin 1983).

Development of mountain valleys for irrigated pasture at the end of the 19th century has been a major factor in the decrease in coldwater habitat within the Humboldt River system. Many smaller tributary streams are completely dewatered by irrigation diversions downstream from national forest boundaries. Other tributaries have become warm and silty as a result of flow depletion, poor-quality irrigation return water, and soil erosion, or they have been channelized and stripped of bank cover (Walstrom 1973; Coffin 1983).

A century of intensive livestock grazing, with its associated stream bank trampling and riparian canopy reduction, has been another major cause of habitat deterioration. This situation has been accentuated along waterways where beavers have eliminated riparian cover and high flows have washed out abandoned beaver dams. An analysis of stream survey data collected from 63 cutthroat trout streams by the NDW in 1980 revealed that 72% of the surveyed waters were only in fair or poor condition (Coffin 1981). More recent surveys by the U.S. Forest Service (USFS) and the U.S. Bureau of Land Management (BLM) indicated a declining trend in stream condition brought about in part by the floods of 1983–1984 (Coffin, personal communication).

The Humboldt River race of Lahontan cutthroat trout appears somewhat more resistant to hybridization with and displacement by nonnative fishes than the Lahontan cutthroat trout of the western Lahontan basin (Behnke 1979). Displacement, however, has eliminated cutthroat trout from two-thirds of the existing coldwater habitat in the drainage (Coffin 1981). In addition, Lahontan cutthroat trout populations are mixed with nonnative trout species, primarily brook trout, in 24 tributaries of the Humboldt River (Coffin 1981). However, as of 1983, only three populations in Humboldt River tributaries were known to be introgressed with rainbow trout, despite periodic plantings of the latter in many streams (Loudenslager and Gall 1980a).

Smoke Creek and Black Rock Subbasins

Once arms of ancient Lake Lahontan, the Black Rock and Smoke Creek deserts contain numerous tributaries suspected of supporting Lahontan cutthroat trout. Surveys performed by the NDW between 1935 and 1960 revealed that 20 streams, largely within the Quinn River watershed were occupied by Lahontan cutthroat trout (Durrant 1935; Frantz and King 1958).

As of 1987, populations of what may be Lahontan cutthroat trout occur in 11 small headwater tributaries of the Quinn River (Coffin, personal communication). However, the genetic purity of these populations remains to be confirmed by electrophoresis. Lahontan cutthroat trout have been reported from a number of other Quinn River tributaries that have not been surveyed. Streamflow depletion, watershed degradation, and hybridization with and displacement by introduced nonnative trout are primarily responsible for the decline of Lahontan cutthroat trout from this subbasin (Coffin, personal communication).

Summit Lake, Nevada situated immediately north of the Black Rock Desert, supports the largest self-sustaining lacustrine population of Lahontan cutthroat trout in existence. The 240-hectare lake has been used as an egg source for the Pyramid Lake cutthroat trout stocking program. Spawning migrations into Mahogany Creek, the sole spawning tributary, have varied from 700 to 5,000 fish annually (Rankel 1976; G. M. Sonnevil, USFWS, personal communication). The origin of the cutthroat trout population in Summit Lake is uncertain. Wheeler (1974) reports that cutthroat trout were transplanted from Pyramid Lake prior to 1880, but others hypothesize that the fish gained natural entrance from Lake Lahontan prior to formation of the present lava flow which isolates the drainage (La Rivers 1962; Rankel 1976).

Honey Lake Drainage

Lahontan cutthroat trout probably occurred in the Honey Lake drainage during historic times. Snyder (1917) collected Lahontan cutthroat trout from the Susan River upstream from Honey Lake in 1915. Although these fish may have originated from plants of Lahontan cutthroat trout made as early as 1904 (Shebley 1904), early settlers reported that cutthroat trout were abundant in the Susan River in 1853, decades before fish stocking was recorded in California waters (Hutchings 1857).

Populations Established Outside the Lahontan Basin

A small number of Lahontan cutthroat trout populations in California have become established outside their native waters as a result of stocking hatchery-reared fish. Between 1893 and 1938, millions of Lahontan cutthroat trout fry, obtained from the eggs of adults trapped in Lake Tahoe's spawning tributaries, were planted in hundreds of lakes and streams throughout California (Behnke 1979). Although most of the naturally maintained populations were displaced by subsequent introductions of other trout, known genetically pure, self-sustaining populations still exist in seven California streams outside the Lahontan Basin.

Nevada Fish Commission reports noted that large numbers of cutthroat trout were hatched, reared, and sent throughout Nevada until about 1930, when the diminishing populations in Pyramid Lake ended this activity (La Rivers 1962). In addition, Miller and Alcorn (1946) reported that ranchers transplanted cutthroat trout from the Reese River drainage to streams in the nearby Toquima Range and on the east slope of the Toiyabe Range. Several of these streams still support Lahontan cutthroat trout (Coffin 1981). A population of Lahontan cutthroat trout that appears to have been transplanted from the Truckee River system was recently discovered in Donner Creek in the Lake Bonneville Basin near the Nevada border (Hickman and Behnke 1979).

It is possible that, as more waters are surveyed, additional populations of Lahontan cutthroat trout established by early stocking activities will be discovered. However, the overall status of the subspecies is not likely to change appreciably.

Life History

In many respects, the life history of Lahontan cutthroat trout resembles that of other subspecies of interior cutthroat trout. (Duff 1988; Liknes and Graham 1988; Varley and Gresswell 1988, all this volume). Lahontan cutthroat trout are obligatory stream spawners; spawning occurs during spring months, generally April through July, depending on stream flow and water temperatures. Spawning migrations typically commence after minimum stream temperatures reach 5°C (Lea 1968; Rankel 1976). Fish from fluvial populations do not usually migrate long distances in search of spawning habitat; however, individuals from lacustrine–adfluvial populations, such as those in Walker Lake, have been reported to migrate as far as 200 km before spawning (Smith and Needham 1935). Fish from fluvial–adfluvial populations of cutthroat trout in the Humboldt River system frequently ascend smaller seasonally intermittent tributaries to spawn (Coffin 1981). Few data on migration time exist. Migrations of radio-tagged adult cutthroat trout in the Truckee River have been measured (USFWS 1979). Daily movements as great as 11 km were observed, although the average movement was only 0.75 km.

Spawning Lahontan cutthroat trout, observed in 12 study streams, prefered gravel sizes ranging from 6 to 50 mm in diameter and water velocities ranging from 4 to 6 cm/s (S. Robertson, USFS, unpublished data). Robertson observed that fry preferred water depths of less than 8 cm and water velocities of less than 15 cm/s. Johnson et al. (1981) noted that fry preferred water depths of 6–43 cm and water velocities of less than 9 cm/s. Water temperatures of less than 13.3°C and intragravel dissolved oxygen levels in excess of 5 mg/L are required during the April through June egg-incubation period (USFWS 1979). Even slight short-duration water temperature increases above 13.3°C during the early portion of the incubation period can result in major mortality (Vigg and Koch 1980). Sustained water temperatures below 6°C during egg incubation are suspected of also being responsible for heavy losses (Hoffman and Scoppettone 1984).

Lacustrine Lahontan cutthroat trout mature at 3–5 years of age, most of them at age 4 (Calhoun 1944a; Lea 1968; Rankel 1976; King 1982). Some fast-growing lacustrine female Lahontan cutthroat trout reared in hatcheries have been known to mature at age 2. At Heenan Lake, most eggs are taken from spawners at ages 4 through 7. In stream environments, male Lahontan cutthroat trout frequently mature at age 2, and females mature at age 3 (Coffin 1981).

Lahontan cutthroat trout fecundity generally increases with increased size of females. Calhoun (1944a) described Lahontan cutthroat trout from upper Blue Lake (Alpine County, California) with fork lengths (FL) of 300, 350, and 400 mm as having 800, 1,100, and 1,700 eggs/fish, respectively. Similarly, Lea (1968) observed 280-, 300-, and 380-mm-FL fish from Independence Lake with 739, 1,200, 2,000 eggs/fish, respectively. By contrast, only 100–300 eggs/fish were observed in 143–175-mm-FL females collected from small Nevada streams (Coffin 1981).

Lahontan cutthroat trout eggs generally hatch in 4–6 weeks depending on water temperatures (Calhoun 1944a; Lea 1968; Rankel 1976). Observations at Blue, Summit, and Independence lakes indicated that progeny of lacustrine–adfluvial spawners generally start emigrating from natal streams shortly after emergence, with migration continuing through the summer. Some juveniles spend one or more years in the tributaries, however, and Rankel (1976) noted that 20% of the juveniles in tributaries to Summit Lake did not migrate to the lake until the following spring. Johnson et al. (1981) conducted an intensive study of juvenile Lahontan cutthroat trout emigration in the Truckee River system and observed that emigration behavior of stocked young of the year was initially determined by juvenile fish density. Summer young-of-the-year mortality was extremely high (up to 95%), and peak emigration occurred during fall and winter freshets.

Stomach analyses of fluvial Lahontan cutthroat trout showed that they are opportunistic feeders whose diets consist of organisms, typically insects, most commonly found in drift (Moyle 1976; Coffin 1981). Lake residents prefer zooplankton, benthic invertebrates, and terrestrial insects when they are available (Calhoun 1942, 1944b; Lea 1968; Rankel 1976; Koch et al. 1979; King 1982). In lakes where Lahontan cutthroat trout have long coexisted with other fishes, larger lake residents, typically those larger than 300 mm (FL), become abpiscivorous (La Rivers 1962; Lea 1968; Koch et al. 1969; King 1982; Sigler et al. 1983).

Lahontan cutthroat trout growth rates are variable. Faster growth occurs in larger, warmer, more fertile waters, particularly those where forage fish are utilized. In such lakes, age-4 Lahontan cutthroat trout typically reach 400–500 mm FL (Table 2). In colder, more oligotrophic lakes such as upper Blue Lake and Independence Lake, age-4 Lahontan cutthroat trout averaged considerably less than 400 mm FL (Calhoun 1944a; Lea 1968). Spawning cutthroat trout (presumably age 4) at relatively infertile Marlette Lake averaged 320 mm FL (King 1982). In Bull Lake, a small alpine lake in the Sierra Nevada, I found that age-4 Lahontan cutthroat trout averaged only 260 mm FL. Growth rates are considerably slower in streams than in lakes. The fish I collected from six Sierra Nevada streams averaged 89, 114, 203, and 267 mm FL at ages 1, 2, 3, and 4, respectively. Similar growth rates have been observed in Nevada streams (Coffin 1983). Resident fish older than age 4 are uncommon in small streams.

The growth potential of Lahontan cutthroat trout, in terms of weight, varies from less than 0.2 kg in small streams to 9 kg or more in large, fertile lakes. Several fish larger than 18 kg were reported from Pyramid Lake prior to 1939 (Behnke 1979).

Historically, Lahontan cutthroat trout were found in a variety of habitats ranging from large alkaline lakes to oligotrophic lakes and streams. Lahontan cutthroat trout, unlike most other *Salmo*, are able to tolerate unusually high levels of alkalinity and dissolved solids, as high as 3,000 mg/L and 10,000 mg/L, respectively (Koch et al. 1979). Fluvial Lahontan cutthroat trout are most abundant in fertile, low-gradient streams, particularly those flowing through meadows.

Lahontan cutthroat trout evolved in the absence of other trout species, and they do not compete well with them (Behnke 1979). In stream environments within the western portion of the Lahontan drainage, Lahontan cutthroat trout have seldom been able to coexist with nonnative

TABLE 2.—Growth of Lahontan cutthroat trout from selected western waters of different relative fertility.

Water, location	Reference	Relative fertility of water	Fork length (mm) at age						
			1	2	3	4	5	6	7
Pyramid Lake, Nevada	Sigler et al. (1983)	High	217	291	362	431	499	573	629
Walker Lake, Nevada	Koch et al. (1979)	High	103	207	318	416	493	559	649
Topaz Lake, California–Nevada	NFGD (1958)	High		215	352	461	533	603	
Heenan Lake, California	Calhoun (1942)	High	83	214	314	455			
Omak Lake, Washington	Kucera et al. (1985)	High	75–150	200–298	350–450	450–550			
Summit Lake, Nevada	Rankel (1976)	High		245	395	462–485	503–533		
Independence Lake, California	Lea (1968)	Low	85	155	218	286	356		
Upper Blue Lake, California	Calhoun (1944a)	Low	63	166	284	347	367		

trout for longer than a decade. Lahontan cutthroat trout, particularly those within the western portion of the Lahontan Basin, also hybridize with rainbow trout (Behnke 1979). Although a few hybridized populations have maintained themselves as introgressed forms (Figure 1) most have been displaced by the more dominant rainbow trout.

Limited coexistence of Lahontan cutthroat trout with nonnative trout has been observed in several tributaries of the Humboldt River downstream from pure headwater populations of Lahontan cutthroat trout. The latter presumably provide a continuous source of cutthroat trout recruitment to lower stream reaches (Coffin 1981).

Management

The Lahontan cutthroat trout was recognized as an endangered species under the 1973 U.S. Endangered Species Act. In 1975, it was reclassified as a threatened species to facilitate its management and to allow angling (Behnke and Zarn 1976).

Both California and Nevada have adopted management programs for maintaining and enhancing existing Lahontan cutthroat trout populations and for establishing new populations within the Lahontan Basin. The major objective of these programs is to increase the abundance of the subspecies and, where possible, each race to levels that will assure their continued existence (Coffin 1983; Gerstung 1986). For management purposes, it is assumed that distinct races of Lahontan cutthroat trout exist in each of the following subbasins: Truckee, Carson, Walker, Humboldt, Reese, and Quinn rivers, and Summit Lake. State management programs will supplement a federal recovery plan to be prepared by the U.S. Fish and Wildlife Service (USFWS).

State management programs consist of the following general elements: (1) inventory, monitoring, and genetic analysis; (2) reestablishment of the subspecies in former habitats; (3) special angling regulations; (4) environmental protection (including land acquisition); (5) habitat improvement; (6) artificial propagation and broodstock maintenance; and (7) population maintenance in lakes.

Inventory, Monitoring, and Genetic Analysis

Personnel from the CDFG, NDW, USFWS, USFS, and BLM have surveyed streams within the historic range of the Lahontan cutthroat trout to determine distribution, relative abundance, and genetic purity of populations and to assess habitat quality. Surveys of tributaries to the Truckee, Carson, and Walker rivers have been completed, and as of 1987, most of the suitable streams within Humboldt River system have been surveyed. Unsurveyed portions of the Humboldt River, primarily tributaries of the East Humboldt River, Little Humboldt River, and Reese River will be surveyed during the late 1980s. During this period, surveys will also be completed on tributaries to the Black Rock Desert–Quinn River system (Coffin, personal communication).

Electrophoretic analyses of samples from all known California populations and from 50% of the Nevada populations have been completed. Some of the findings have been published (Loudenslager 1979; Loudenslager and Gall 1980a, 1980b, 1980c). Analyses of meristic characters have been completed for samples from 12 additional Nevada stream populations (Behnke 1960).

Periodically, population and habitat trend data have been and will continue to be collected from California and Nevada Lahontan cutthroat trout streams by state and federal agencies as part of the current management program (Coffin 1983; Gerstung 1986).

Population Establishment and Reestablishment

The NDW management plan for Lahontan cutthroat trout recommends that Lahontan cutthroat trout be stocked in suitable fishless streams, including those streams where trout populations have been eliminated by drought or flash floods. Twelve streams that fit this category have been identified for stocking (Coffin 1983). In addition, approximately 30 other Nevada streams, currently occupied by nonnative trout, have been identified for chemical treatment and restocking with Lahontan cutthroat trout (Coffin 1983).

The California cutthroat trout management plan recommends chemical treatment and subsequent stocking of Lahontan cutthroat trout in at least 10 streams in the Lahontan drainage (Gerstung 1986). Both the California and Nevada plans recommend the selection of easily treatable waters in areas of low public use. Lahontan cutthroat races endemic to the subbasin will be used for restocking when available. By 1987, only a small number of California and Nevada streams had been repopulated with Lahontan cutthroat trout through the chemical treatment and restocking programs.

Special Angling Regulations

Lahontan cutthroat trout, like most cutthroat trout subspecies, are generally very vulnerable to depletion by angling (Behnke and Zarn 1976). As a result, reduced bag limits and increased angling closures may be required in areas of heavy recreational use. Of 17 California waters currently occupied by Lahontan cutthroat trout, 7 are closed to angling and 3 are restricted to catch-and-release angling.

In Nevada, special regulations consisting of reduced bag limits and size limits have been in effect many years on Walker and Pyramid lakes and will likely, with periodic modification, be continued. Summit Lake is closed to public fishing to protect the brood stock used for stocking Pyramid Lake and to reserve fishing for the Indian tribe that owns the lake (Rankel 1976). Angling closures and special regulations have not been established on Lahontan cutthroat trout streams in Nevada because they are generally subject to low angler use.

In the future, more special regulations, particularly those encouraging catch-and-release angling, are likely to be enacted for California and Nevada waters as an alternative to angling closures or artificial stocking. Catch-and-release regulations have been very effective in western cutthroat trout waters (Behnke and Zarn 1976; Gresswell and Varley 1988, this volume).

Environmental Protection and Habitat Improvement

The CDFG and NDW routinely review and comment on proposed projects that could adversely affect stream habitats. If a federally listed threatened or endangered species, such as the Lahontan cutthroat trout, is present, the Endangered Species Act requires a consultation process with the USFWS.

Most existing and proposed Lahontan cutthroat trout waters in California are situated in national forest wilderness or in relatively undeveloped areas where damaging habitat alteration is unlikely to occur. In contrast, Nevada populations are much more likely to be adversely affected by resource development. As noted before, water development has impaired or eliminated many former Lahontan cutthroat waters, and some additional losses are expected.

Surface mining in Nevada is one of the biggest threats to the Lahontan cutthroat trout because much of the state is mineralized, and the escalating price of some minerals is stimulating an increase in mining activities. Pollution and siltation associated with large-scale surface mining can devastate a small cutthroat trout stream if effective protective measures are not taken. Some progress is being made in mitigating mining damage (Coffin, personal communication).

In both California and Nevada, public land holdings along Lahontan cutthroat trout streams are often intermingled with private parcels. Because this fragmented ownership pattern complicates stream habitat protection and management, Nevada and California management plans will encourage public acquisition of key parcels of privately owned land (Coffin 1983; Gerstung 1986). However, funding is limited, and land acquisitions may be limited to those accomplished by land exchanges, primarily those resulting from ongoing BLM and USFS land-consolidation programs. Both these agencies are being encouraged to give priority to acquiring parcels threatened by incompatible development. The CDFG has acquired parcels of privately owned land along several streams that support Lahontan cutthroat trout and has proposed acquiring private inholdings along several others (Gerstung 1986).

Stream-bank and watershed damage from heavy livestock grazing continues to be a major problem in Nevada. Recent studies of western trout streams have shown that dramatic improvements in stream habitat quality and trout abundance occur after stream banks are fenced or stabilized, riparian vegetation is restored, and cover and pool-forming devices are installed (Platts and Renni 1985). Within Nevada, habitat improvement projects have been initiated on Mahogany Creek, the major spawning area for Summit Lake, and on Sherman and Frazier creeks within the Humboldt River drainage (Coffin 1983). The NDW and USFS are collecting long-term information on Lahontan cutthroat trout population and habitat quality trends from Gance Creek within the Humboldt River drainage, where the effects of cattle exclosure are being evaluated as part of the Saval Ranch Research Project (Coffin 1981; Platts and Nelson 1982). Major range and riparian restoration projects, which will probably continue through the 1990s, are being initiated within the North Fork Humboldt, Little Humboldt, and Marys river drainages (Coffin 1983).

In California, less degradation of cutthroat trout habitat has occurred; hence there has been less emphasis on habitat improvement. Nevertheless, five stream habitat improvement projects involving bank stabilization and livestock exclosures have been completed and several more are planned (Gerstung 1986).

Artificially Maintained Populations

From 1883 to 1938, Lake Tahoe and hundreds of other California lakes and streams were planted annually with Lahontan cutthroat trout fingerlings. Although accounts of good fishing exist in old records, the effectiveness of the planting program was never evaluated. Since 1940, Lahontan cutthroat trout fingerling stocking within California has been limited to approximately 30 coldwater reservoirs and lakes. Limited evaluation indicates that fair to good angling is being maintained in several high-elevation coldwater lakes where the cutthroat trout are stocked in the absence of other fish. The best angling, however, has occurred in the more fertile midelevation lakes such as those in eastern California. For example, fertile Martis Creek Reservoir and Heenan Lake, both stocked annually with yearling cutthroat trout and managed for catch-and-release angling, each produced peak landings of 6,000 Lahontan cutthroat trout weighing 0.5-1.5 kg (Deinstadt 1982, in press). Martis Lake, however, is no longer managed for Lahontan cutthroat trout.

In contrast, Lahontan cutthroat trout stocked in Independence Lake, an oligotrophic water with an abundance of nongame fish and introduced kokanee *Oncorhynchus nerka kennerly*, have not fared well. Heavy stocking during a 10-year evaluation period failed to improve angling or spawning escapement (Gerstung 1986). It is suspected that the hatchery-reared Lahontan cutthroat trout were unable to compete with the abundant native nongame fish and kokanee.

Lahontan cutthroat trout fingerlings have been planted in several coldwater lakes and reservoirs in Nevada, with varied results. A notable fishery for trophy-sized cutthroat trout existed at lakes Mead and Mojave on the Colorado River until striped bass *Morone saxatilis* became established. Lahontan cutthroat trout fingerlings planted in high-elevation lakes in the Ruby Mountains provide a limited backcountry cutthroat trout fishery (Coffin 1983, and personal communication).

Most lake management effort in Nevada is directed toward Pyramid and Walker lakes. Pyramid Lake now supports a popular sport fishery for Lahontan cutthroat trout that is maintained by stocking 1–3 million cutthroat trout annually. Coleman and Johnson (1988, this volume) report that the cutthroat trout fishery supports an annual angler use of 35,000 to 141,000 angler-days and an annual harvest of up to 84,000 Lahontan cutthroat trout. However, the future of the Pyramid Lake fishery is uncertain because the declining lake level and the increasing concentrations of total dissolved solids could eventually eliminate the Lahontan cutthroat trout fishery (Sigler et al. 1983). The Pyramid Lake fishery is described in greater detail by Coleman and Johnson (1988).

Walker Lake supports a popular sport fishery for Lahontan cutthroat trout that is maintained by stocking an average of 250,000 yearlings annually. The quality of the fishery is strongly influenced by climatic cycles. For example, during the last dry cycle (between 1971 and 1981), the level of Walker Lake dropped by 7.4 m, and the annual Lahontan cutthroat trout harvest declined from a peak of 47,000 fish in 1963 to a low of 3,500 fish in 1976 (Sevon 1982, 1987). Following several seasons of exceptionally heavy runoff the level of Walker Lake rose (6 m by 1987), and cutthroat trout production substantially improved. The estimated annual harvests in 1985 and 1986 exceeded 50,000 Lahontan cutthroat trout (Sevon 1987). The majority of captured cutthroat trout are in their second or third year of lake residence and average 400 mm FL (Sevon 1987).

The artificially maintained Lahontan cutthroat trout fisheries occurring in California and Nevada lakes are supported by stocking fish produced from eggs obtained at Heenan Lake in California and from Marlette, Summit, Catnip and Pyramid lakes in Nevada. At least 5 million eggs are reported to be available annually from these sources (King 1982; Sevon 1987; Coleman and Johnson 1988).

Acknowledgments

I am grateful to Patrick D. Coffin of the Nevada Department of Wildlife for his critical review of this manuscript. Don Sada of the U.S. Fish and Wildlife Service, Don Duff of the U.S. Forest Service, and Steven Nicola of the California Department of Fish and Game provided many helpful comments. I am particularly indebted to the late Dale Lockhard, formerly of the Nevada Department of Wildlife and to the late David Koch, formerly of the Desert Research Institute for valuable advice and support.

References

Allan, R. 1958. Fisheries management report–Walker Lake. Nevada Fish and Game Department, Reno.
Bailey, R. 1978. Restoration of a wild Lahontan cut-

throat trout fishery in the Truckee River, California–Nevada. Pages 53-55 *in* J. R. Moring, editor. Proceedings of wild trout–catchable trout symposium. Oregon Department of Fish and Wildlife, Corvallis.

Behnke, R. J. 1960. Taxomony of the cutthroat trout of the Great Basin. Master's thesis. University of California, Berkeley.

Behnke, R. J. 1974. The effects of the Newlands project on the Pyramid Lake fishery. Colorado State University, Fort Collins.

Behnke, R. J. 1979. The native trouts of the genus *Salmo* of western North America. Report to U.S. Fish and Wildlife Service, Denver, Colorado.

Behnke, R. J. 1988. Phylogeny and classification of cutthroat trout. American Fisheries Society Symposium 4:1–7.

Behnke, R. J., and M. Zarn. 1976. Biology and management of threatened and endangered western trout. U.S. Forest Service General Technical Report RM-28.

Benson, L. V. 1978. Fluctuations in the level of pluvial Lake Lahontan for the past 40,000 years. Quarternary Research 9:300–318.

Calhoun, A. J. 1942. The biology of the black-spotted trout (*Salmo clarkii henshawii*) (Gill and Jordan) in two Sierra Nevada lakes. Doctoral desertation. Stanford University, Palo Alto, California.

Calhoun, A. J. 1944a. Black-spotted trout in Blue Lake, California. California Fish and Game 30:22–42.

Calhoun, A. J. 1944b. The food of the black-spotted trout (*Salmo clarkii henshawi*) in two Sierra Nevada lakes. California Fish and Game 30:80–85.

Coffin, P. D. 1981. Distribution and life history of the Lahontan/Humboldt cutthroat trout, Humboldt River drainage basin. Nevada Department of Wildlife, Reno.

Coffin, P. D. 1983. Lahontan cutthroat trout fishery management plan for the Humboldt River drainage basin. Nevada Department of Wildlife, Federal Aid in Fish Restoration, Project F-20-17, Reno.

Coleman, M. E., and V. K. Johnson. 1988. Summary of trout management at Pyramid Lake, Nevada, with emphasis on Lahontan cutthroat trout, 1954–1987. American Fisheries Society Symposium 4:107–115.

Cordone, A. J., and T. C. Frantz. 1966. The Lake Tahoe sport fishery. California Fish and Game 52: 240–274.

Cordone, A. J., and T. C. Frantz. 1968. An evaluation of trout planting in Lake Tahoe. California Fish and Game 54:68–69.

Curtis, B. 1938. Proposed management program for Lake Tahoe fishery based on investigations made in 1938. California Division of Fish and Game, Sacramento.

Dadd, B. 1869. Great trans-continental railroad guide. George A. Crofutt, Chicago.

Deinstadt, J. 1982. The Martis Reservoir trout fishery, Placer County, California. California Department of Fish and Game, Inland Fisheries Division, File report, Sacramento.

Deinstadt, J. M. In press. California's use of catch and release angling regulations on trout waters. *In* R. A. Barnhart and T. D. Roelofs, editors. Catch-and-release fishing: a decade of experience. Humboldt State University, Arcata, California.

Duff, D. A. 1988. Bonneville cutthroat trout: current status and management. American Fisheries Society Symposium 4:121–127.

Durrant, S. D. 1935. A survey of the waters of the Humboldt National Forest, Nevada. U.S. Bureau of Fisheries, Washington, D.C.

Evermann, B. W., 1906. The golden trout of the southern high Sierras. U.S. Bureau of Fisheries Bulletin 25:1–51.

Evermann, B. W., and H. C. Bryant. 1919. California trout. California Fish and Game 3:105–135.

Frantz, T. C., and D. J. King. 1958. Stream–lake survey, 13. Nevada Fish and Game Commission, Completion Report, Reno.

Fremont, J. C. 1845. Report of the exploring expeditions to the Rocky Mountains in the year 1842, and to Oregon and North California in the years 1843-1844. U.S. Senate, 28th Congress, 2nd Session, Executive Document 174. (Printed by Gale and Seaton, Washington, D.C.)

Gerstung, E. R. 1986. Fishery management plan for Lahontan cutthroat in California and Nevada waters. California Department of Fish and Game, Inland Fisheries Branch, Sacramento.

Gresswell, R. E., and J. D. Varley. 1988. Effects of a century of human influence on the cutthroat trout of Yellowstone Lake. American Fisheries Society Symposium 4:45–52.

Hickman, T. J., and R. J. Behnke. 1979. Probable discovery of the original Pyramid Lake cutthroat trout, Progressive Fish-Culturist 41:135–137.

Hoffman, R. J., and G. G. Scoppettone. 1984. Effect of water quality on survival of Lahontan cutthroat trout eggs in the Truckee River, west-central Nevada and eastern California. U.S. Geological Survey Open-file Report 84-437, Carson City, Nevada.

Hunt, E. W. 1910. Twenty-first biennial report of the Board of Fish and Game Commissioners of the State of California for the years 1909–1910. Superintendent of State Printing, Sacramento.

Hutchings, J. M. 1857. A jaunt to Honey Lake Valley and Noble's Pass. Hutchins Illustrated California Magazine 1(12):317–329. (Reprint 1962. Howell North, Berkeley, California.)

James, G. W. 1921. The lake of the sky, Lake Tahoe. Radiant Life Press, Pasadena, California.

Johnson, E. C. 1974. Walker River Paiutes, a tribal history. University of Utah Printing Service, Salt Lake City.

Johnson, G. L., D. H. Bennett, and T. C. Bjornn. 1981. Juvenile emigration of Lahontan cutthroat trout in the Truckee River–Pyramid Lake System. University of Idaho, College of Forest, Wildlife and Range Sciences, Moscow.

Jordan, D. S., and H. W. Henshaw. 1878. Report upon the fishes collected during the years 1875, 1876, and 1877, in California and Nevada. Pages 187-200 *in* Annual report United States Geological Survey, west of 100th Meridian (Wheeler Survey). U.S.

Geological Survey, Annual Report of Chief of Engineers for 1878, Part 3, Appendix NN, Washington, D.C.

Juday, C. 1907. Notes on Lake Tahoe, its trout and trout fishing. U.S. Bureau of Fisheries Bulletin 26:133–146.

King, J. W. 1982. Investigation of the Lahontan cutthroat trout broodstock at Marlette Lake, Nevada. Master's thesis. University of Nevada, Reno.

Koch, D. L., J. J. Cooper, E. L. Lider, R. L. Jacobson, and R. J. Spencer. 1979. Investigations of Walker Lake, Nevada: dynamic ecological relationships. University of Nevada, Desert Research Center, Reno.

Koch, D. L., and seven coauthors. 1977. Proposal to investigate the feasibility of altering the chemical balance of Walker Lake water with biological benefits to the lake, employment of people, and marketable by-projects. University of Nevada, Desert Research Institute, Reno.

Kucera, P. A., D. L. Koch, and G. F. Marco. 1985. Introductions of Lahontan cutthroat trout into Omak Lake, Washington. North American Journal of Fisheries Management 5:296–301.

La Rivers, I. 1962. Fishes and fisheries of Nevada. Nevada State Fish and Game Commission, Reno.

Lea, T. N. 1968. Ecology of the Lahontan cutthroat trout, *Salmo clarkii henshawi*, in Independence Lake, California. Master's thesis. University of California, Berkeley.

Liknes, G. A., and P. J. Graham. 1988. Westslope cutthroat trout in Montana: life history, status, and management. American Fisheries Society Symposium 4:53–60.

Loudenslager, E. J. 1979. Biochemical and cytogenetic systematics of inland cutthroat subspecies. University of Wyoming, National Park Service Research Center, Laramie.

Loudenslager, E. J., and G. A. E. Gall. 1980a. Biochemical systematics of Nevada trout populations. University of California, Davis.

Loudenslager, E. J., and G. A. E. Gall. 1980b. Cutthroat trout, a biochemical-genetic assessment of their status and systematics. University of California, Davis.

Loudenslager, E. J., and G. A. E. Gall 1980c. Geographic patterns of protein variation and subspeciation in cutthroat trout, *Salmo clarki*. Systematic Zoology 29:27–42.

McAfee, W. R. 1966. Lahontan cutthroat trout. Pages 225-231 *in* A. Calhoun, editor. Inland fisheries management. California Department of Fish and Game, Sacramento.

Miller, R. R. 1951. The natural history of Lake Tahoe fishes. Doctoral dissertation. Stanford University, Palo Alto, California.

Miller, R. R., and J. R. Alcorn. 1945. The introduced fishes of Nevada, with a history of their introduction. Transactions of the American Fisheries Society 73:173–193.

Moyle, P. B. 1976. Inland fishes of California. University of California Press, Berkeley.

NFGD (Nevada Fish and Game Department). 1958. Fisheries management report. Tahoe and Topaz lakes. Federal Aid in Fish Restoration, Project FAF-4-R, Job Completion Report, Reno.

Platts, W. S., and R. D. Nelson. 1982. Livestock fishery interaction studies—Gance Creek, Nevada. U.S. Forest Service, Intermountain Forest and Range Experiment Station, Progress Report 4, Boise, Idaho.

Platts, W. S., and J. N. Renni. 1985. Riparian and stream enhancement and research in the Rocky Mountains. North American Journal of Fisheries Management 5:115–125.

Rankel, G. L. 1976. Fishery management program, Summit Lake Indian Reservation, Humboldt County, Nevada. U.S. Fish and Wildlife Service, Division of Fishery Services, Special Report, Reno.

Rutter, C. 1902. Notes on fishes from streams and lakes of northeastern California not tributary to the Sacramento basin. United States Fish Commission Bulletin 22:145–148.

Scott, E. B. 1957. The saga of Lake Tahoe; a complete documentation of Lake Tahoe's development over the last one hundred years. Sierra-Tahoe Publishing, Crystal Bay, Nevada.

Sevon, M. 1982. Fisheries management report, Walker Lake 1981. Nevada Department of Wildlife, Federal Aid in Fish Restoration, F-20-17, Job Completion Report, Reno.

Sevon, M. 1987. Fisheries management report, Walker Lake 1986. Nevada Department of Wildlife, Federal Aid in Fish Restoration, F-20-22, Job Completion Report, Reno.

Shebley, W. H. 1894. Thirteenth biennial report of the State Board of Fish Commissioners of the State of California for the years 1893–1894. Superintendent of State Printing, Sacramento.

Shebley, W. H. 1904. Eighteenth biennial report of the State Board of Fish Comissioners of the State of California for the years 1903–1904. Superintendent of State Printing, Sacramento.

Shebley, W. H. 1929. History of the fish and fishing conditions of Lake Tahoe. California Fish and Game 15:194–203.

Sigler, W. F., W. T. Helm, P. A. Kucera, S. Vigg, and G. W. Workman. 1983. Life history of the Lahontan cutthroat trout, *Salmo clarki henshawi*, in Pyramid Lake, Nevada. Great Basin Naturalist 43:1–29.

Smith, H. M. 1898. Report of the Commissioner for the year ending June 30, 1886. U.S. Commission of Fish and Fisheries 22:573.

Smith, O. R., and P. R. Needham. 1935. A stream survey in the Mono and Inyo national forests, California, 1934. U.S. Bureau of Fisheries, Washington, D.C.

Snyder, J. O. 1917. The fishes of the Lahontan system of Nevada and northeastern California. U.S. Bureau of Fisheries Bulletin 35:31–86.

Sumner, F. H. 1939. The decline of Pyramid Lake fishery. Transactions of the American Fisheries Society 69:215–224.

Townley, J. M. 1980. The Truckee basin fishery, 1844–1944. Nevada Historical Society in cooperation with the Desert Research Institute, University of Nevada, Reno.

Trelease, T. J. 1952. The death of a lake. Field and Stream 56:30–31, 109–111.

USFWS (U.S. Fish and Wildlife Service). 1979. Restoration of a reproductive population of Lahontan cutthroat trout (*Salmo clarki henshawi*) to the Truckee River/Pyramid Lake System. Fisheries Assistance Office, Special Report, Reno, Nevada.

Varley, J. D., and R. E. Gresswell. 1988. Ecology, status, and management of the Yellowstone cutthroat trout. American Fisheries Society Symposium 4:13–24.

Vestal, E. H. 1950. Chemical treatment of Upper Twin Lake, Robinson Creek, Mono County, California. California Division of Fish and Game, Bureau of Fish Conservation, Sacramento.

Vigg, S. C., and D. L. Koch. 1980. Upper lethal temperature range of Lahontan cutthroat trout in waters of different ionic concentration. Transactions of the American Fisheries Society 109:336–339.

Walstrom, R. E. 1973. Forecasts for the future—fish and wildlife: water for Nevada, number 6, appendix D. Inventory: statistical data for streams and lakes of Nevada. Nevada State Water Engineer's Office, Carson City.

Wedertz, F. S. 1978. Mono diggings. Chalfant Press, Bishop, California.

Welch, W. R. 1929. Trout fishing in California today and fifty years ago. California Fish and Game 15:20–22.

Wheeler, S. S. 1974. The desert lake. Caxton Printers, Caldwell, Idaho.

American Fisheries Society Symposium 4:107–115, 1988

Summary of Trout Management at Pyramid Lake, Nevada, with Emphasis on Lahontan Cutthroat Trout, 1954–1987

MARK E. COLEMAN[1]

U.S. Fish and Wildlife Service, National Fishery Research Center, Building 204
Naval Station Puget Sound, Seattle, Washington 98115, USA

V. KAY JOHNSON[2]

Nevada Department of Wildlife, 1100 Valley Road, Reno, Nevada 89520, USA

Abstract.—Following the extinction of the native strain of Lahontan cutthroat trout *Salmo clarki henshawi* in Pyramid Lake about 1944, trout stocking began in 1954 to restore the sport fishery at the lake. From 1954 to 1964, a trout fishery primarily for rainbow trout *Salmo gairderni*, with creel rates averaging 0.10 fish/h, supported 20,000–40,000 angler-days of effort annually. Harvest ranged from 4,600 to 24,000 fish/year; mean size of creeled fish varied from 358 to 424 mm annually. Fish over 600 mm long were rare in creels. Lahontan cutthroat trout stocking dominated after 1964, the year trophy fishing regulations (483-mm minimum-size limit, flies and lures only, three-fish bag limit) were implemented. Overall landing rates have varied from 0.08 to 0.32 fish/h (mean, 0.20) since 1964. Creel rates varied from 0.02 to 0.14 fish/h and were closely related to size-limit regulations. No strong correlation occurred between landing rate and stocking level. The Lahontan cutthroat trout fishery has sustained an annual effort of 35,000–141,000 angler-days; harvest has ranged from 2,600 to 84,000 fish. The mean size of creeled cutthroat trout varied from 500 to 607 mm, fluctuating with changes in the minimum size limit. Lahontan cutthroat trout over 600 mm long made up 2–55% of the annual harvest (mean, 25%). Anglers have harvested only 1.7% of the Lahontan cutthroat trout stocked since 1975. This low return, combined with the low landing rates, suggests poor survival of fish to harvestable size. None of the strains of Lahontan cutthroat trout available appear to be well suited for Pyramid Lake. Development of a Pyramid Lake strain from adult survivors is recommended.

Pyramid Lake, Nevada, formerly supported a population of Lahontan cutthroat *Salmo clarki henshawi* that attained a size greater than any other trout native to western North America (Behnke and Zarn 1976). The official world record cutthroat trout, weighing 18.6 kg, was caught there in 1925, and there were unconfirmed reports of cutthroat trout over 28 kg taken in the early 1900s by commercial fishermen (Wheeler 1974). Cutthroat trout in the last spawning run in the Truckee River in 1938 had an average weight of 9 kg (Sumner 1940).

Large cutthroat trout were in such abundance that they supported a sizable commercial and a world-famous sport fishery in the late 19th and early 20th centuries. Records indicated that between 90,000 and 120,000 kg were shipped annually by rail car to commercial markets in the region, and the fish were also consumed by Paiute Indian communities on the lake.

[1]Present address: U.S. Fish and Wildlife Service, HC 30 Box 78B, Chiloquin, Oregon 97624, USA.
[2]Retired.

The native strain of Lahontan cutthroat trout in Pyramid Lake evolved its specialized lacustrine attributes of longevity, large size, and piscivory over thousands of years in a continuous lake environment (Behnke and Zarn 1976). The cutthroat trout was the top predator in a vast inland sea (Lake Lahontan) where forage fish, principally tui chub *Gila bicolor*, abounded. Pyramid Lake is a remnant of Lake Lahontan, a pluvial lake that once occupied 13,000 km^2 of present-day northwestern Nevada and northeastern California. Pyramid Lake is the only remnant of the ancient lake that has been inhabited by fish continuously since Lake Lahontan receded (Benson 1978).

The abundant fish fauna in Pyramid Lake and adjacent Winnemucca Lake attracted humans to the area at least 4,000 years ago (Knack and Stewart 1984). Archeological evidence indicates that the present Paiute Indians lived successfully at Pyramid Lake for the past 600 years, using the abundant fishes with no apparent decline in the fish populations (Wheeler 1974). It was not until Europeans settled in the basin that problems developed.

The chronology of the decline and eventual extinction of the native Pyramid Lake strain has been reported previously (Sumner 1940; Trelease 1969a, 1969b; Wheeler 1974; Knack and Stewart 1984). Townley (1980) provided a very thorough analysis of the demise of the fishery from 1844 until extinction, about 1944. All authors agreed that the single most detrimental impact was the Newlands Project of the U.S. Bureau of Reclamation, which began in 1905. Derby Dam, constructed approximately 64 km above the lake, blocked access to the Truckee River upstream, resulting in an 85% decrease in the potential trout spawning and rearing area (R. J. Behnke, Colorado State University, unpublished). Huge transbasin diversions of water resulted in an annual flow loss in the Truckee River of approximately 50%. In years of low precipitation, virtually the entire river was diverted. Between 1905 and 1981, the level of Pyramid Lake fell 26 m. The falling lake level created very shallow water over the delta at the mouth of the Truckee River that during low-water years, restricted fish access to spawning areas. Passage problems were noted as early as 1911 (Snyder 1917). Other factors contributing to the decline of the fishery included water pollution, sedimentation, introductions of exotic species, commercial fishing, and construction of numerous small irrigation dams and diversions.

Following the extinction of the Pyramid Lake strain of Lahontan cutthroat trout, studies were conducted in the 1940s to determine whether any trout could survive the increased total dissolved solids concentration observed in the lake (La Rivers 1962). High survival of cutthroat trout in bioassay and live-cage studies led to renewed interest in restoring a Pyramid Lake trout fishery. The key event leading to the creation of a trout fishery was the accidental catch of numerous large rainbow trout *Salmo gairdneri* in Pyramid Lake during January 1953 (Trelease 1969b). Rainbow trout stocking the next year marked the start of the rebirth of the Pyramid Lake fishery.

It is assumed that natural reproduction of trout in the lower Truckee River has been negligible since the fishery returned. Until 1976, the Truckee River was generally inaccessable to trout due to the delta. In that year, the Marble Bluff Fish Facility began operation. Its canal and ladder system were designed to bypass the delta and allow access to upstream spawning areas (Scoppettone et al. 1986). However, habitat conditions in the lower Truckee River have been considered inadequate for successful reproduction (USFWS

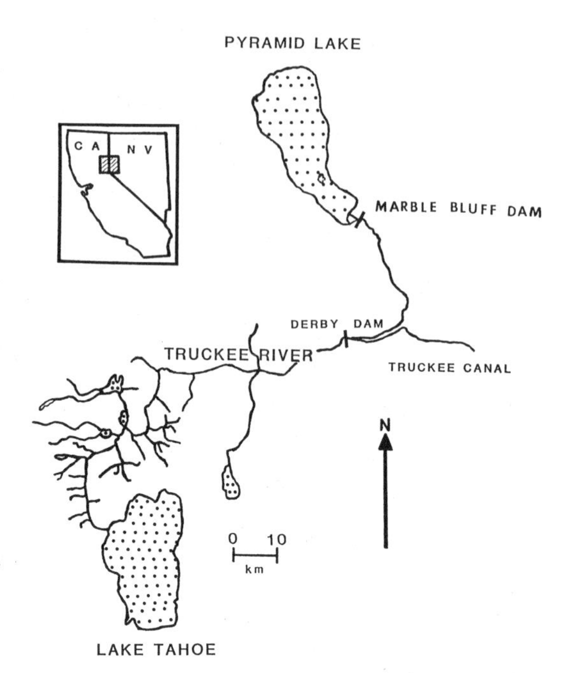

FIGURE 1.—Map of the Pyramid Lake and Truckee River study area. CA = California; NV = Nevada.

1983). Late spring and summer water temperatures frequently exceed 20°C, and intergravel dissolved oxygen levels are low.

The purpose of this paper is to summarize the Pyramid Lake trout fishery and its management from 1954 to 1987. Stocking programs, catch statistics, and management changes are described and analyzed. Emphasis is given to the Lahontan cutthroat trout fishery after 1963.

Study Area

Pyramid Lake is a large, deep graben lake located 45 km northeast of Reno, Nevada, entirely within the Pyramid Lake Paiute Indian Reservation (Figure 1). This north–south oriented lake had a surface area of 450 km^2, a maximum depth of 110 m, and mean depth of 60 m in 1987. It is 40 km long and 6.5–16 km wide. The lake is the terminus of the Truckee River system, which originates 193 river kilometers upstream at Lake Tahoe. The Pyramid Lake Basin, located in the rain shadow of the Sierra Nevada Mountains, has a characteristic high desert climate with an average annual precipitation of 15 cm; evaporation is about 120 cm annually.

Pyramid Lake water is alkaline and saline, composed chiefly of sodium salts of chloride, bicarbonate, and carbonate. The (decreasing)

order of cation abundance is Na^+, K^+, Mg^{2+}, Ca^{2+}. Total dissolved solids concentration is 4,950 mg/L, and pH is 9.2. The lake's surface elevation is 1,165 m. Surface temperatures range from 6° to 23°C, and thermal stratification occurs between May and December (Galat et al. 1981); the lake is classified as a warm monomictic lake (Wetzel 1978). Dissolved oxygen concentrations are near saturation in the surface waters; hypolimnetic concentrations are slightly lower.

The lake has a low species diversity of zooplankton, phytoplankton, benthos, and periphyton (Galat et al. 1981); however, densities of organisms are indicative of a mesotrophic lake (Cole 1979). Five fish species are present in the following order of numerical abundance: tui chub, Tahoe sucker *Catostomus tahoensis*, Lahontan cutthroat trout, cui-ui *Chasmistes cujus*, and Sacramento perch *Archoplites interruptus*. The tui chub numerically accounts for over 70% of the fish in the lake (Vigg 1981).

Trout Stocking

Trout stocking in the Pyramid Lake–lower Truckee River system essentially began in 1954. As state and federal hatcheries redirected and expanded their production towards restoring the trout fishery in Pyramid Lake, the weight, and to a lesser extent the numbers, of fish stocked increased dramatically during the early 1950s (Figure 2). Stocking leveled out at about 10,000 kg in 1958 and hovered between 7,000 and 15,000 kg (45,000–265,000 fish greater than 125 mm) until 1972. A major increase in stocking levels that occurred in 1973 was related to continued improvement in Lahontan cutthroat trout culture techniques and expansion of hatchery facilities. Annual releases from 1973 to 1986 ranged between 0.8 and 2.9 million trout (18,000–53,000 kg).

During the 1950s and 1960s, several species of salmonids were released on an experimental basis. They included brown trout *Salmo trutta*, brook trout *Salvelinus fontinalus*, coho salmon *Oncorhynchus kisutch*, steelhead (anadromous *Salmo gairdneri*), Kamloops rainbow trout, Yellowstone cutthroat trout *Salmo clarki bouvieri*, kokanee *Onchorhynchus nerka*, and rainbow trout × Lahontan cutthroat trout hybrids. From 1954 to 1964, rainbow trout stocking predominated, accounting for over 75% of the fish stocked. Approximately 45,000–175,000 rainbow trout over 125 mm were stocked annually. A few fry (fish <50 mm long) and fingerling (50–125 mm long) plants were made in the 1950s but these were discontinued in 1960.

In most cases, stockings of nonnative trout were failures. These fish apparently were not able to adjust to the highly alkaline waters. Live-cage and bioassay tests conducted at various times indicated that even rainbow trout were not well suited to the unique water chemistry of the lake. In almost all tests, fish showed visible signs of stress when placed in lake water, and many of them died. Alkalinity stress was less acute if fish were tempered over several hours, but even so some fish died. It appears likely that, although many rainbow trout could acclimate to Pyramid Lake water, they were preyed on by birds and large trout upon release. In contrast, tests conducted on Lahontan cutthroat trout indicated these fish readily acclimated to the water; signs of stress measured by behavioral and body pigmentation changes were seldom observed and mortality was rare (Knoll et al. 1979).

The Lahontan cutthroat trout stocking program essentially began in 1961 (Table 1). Sporatic releases occurred as early as 1950, but they consisted mostly of fry or small numbers of larger fish. It was not until 1965 that the Lahontan cutthroat trout became the major fish species reared. Afterwards, this species made up over 75% of the fish stocked; from 1979 to the present it has been the only species released.

Several sources of Lahontan cutthroat trout have been used for propagation programs. From 1954 to 1975, most cutthroat trout were Heenan Lake strain. Brood stock for this strain were maintained at Heenan Lake, California, and Catnip and Marlette lakes, Nevada. A small number of Walker Lake strain were also stocked from Marlette and Catnip lakes. The Summit Lake strain was the dominant fish stocked in the late 1970s and 1980s. Eggs were taken from fish spawned at Summit Lake, Nevada, and from hatchery brood stock held at Lahontan National Fish Hatchery (NFH). In the early 1980s, Independence Lake strain Lahontan cutthroat from Heenan Lake, California, were reared and released in small numbers. Since 1982, increasing numbers of eggs have been taken from fish surviving in Pyramid Lake. In 1987, all eggs for hatchery rearing were taken from spawners at Pyramid Lake. These fish are predominantly of Summit Lake ancestry.

Trout stocked in Pyramid Lake have been reared at several state, federal, and tribal hatcheries. From 1953 to 1975, most production oc-

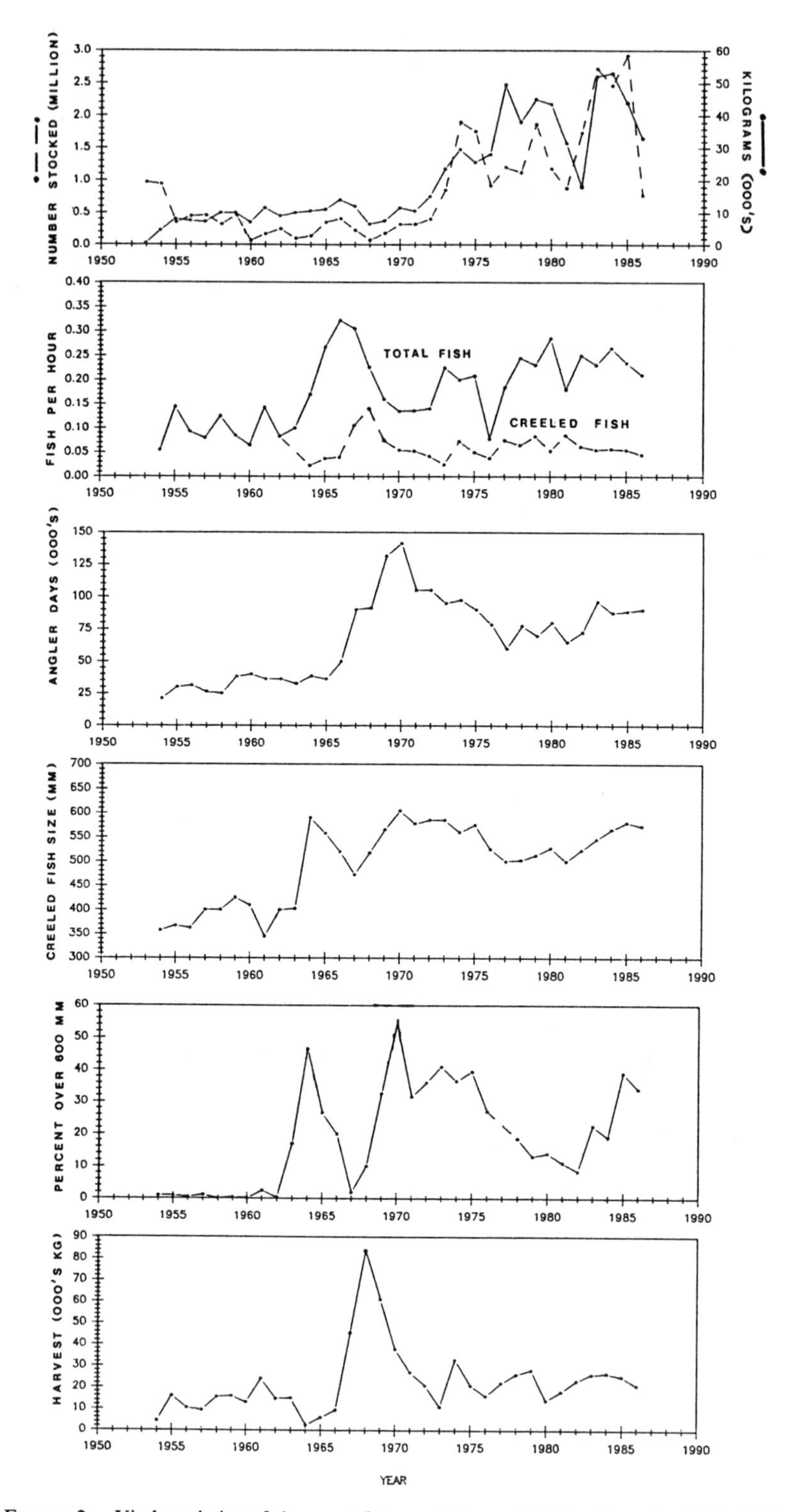

FIGURE 2.—Vital statistics of the sport fishery at Pyramid Lake, Nevada, 1953–1986.

TABLE 1.—Stocking rates of Lahontan cutthroat trout by numbers and weight in Pyramid Lake, Nevada, 1961–1985. Fry and fingerlings were 25–125 mm long and 2–8 months old; subcatchables were fish greater than 125 mm long and 1–2 years old.

Year	Numbers of fish stocked: Fry and fingerlings	Numbers of fish stocked: Subcatchables	Total weight (kg)
1961	10,403	59,880	2,307
1962	112,599	50,725	2,262
1963	16,674	19,251	1,699
1964		53,984	2,242
1965	112,355	129,713	11,128
1966	194,000	206,664	13,896
1967	20,315	168,333	12,218
1968		81,325	6,767
1969	43,427	124,926	7,426
1970	157,125	89,088	11,706
1971	122,560	167,580	10,665
1972	116,000	265,625	15,050
1973	455,086	359,234	23,314
1974	1,137,867	735,556	34,782
1975	1,422,440	333,701	25,497
1976	413,940	499,880	27,891
1977	514,981	618,519	49,231
1978	517,316	523,276	36,900
1979	1,271,831	771,062	45,209
1980	424,317	588,400	43,789
1981	316,277	568,858	31,471
1982	1,497,885	241,253	17,952
1983	1,992,363	723,894	52,079
1984	1,544,304	668,650	53,387
1985	2,225,847	684,410	44,388

curred at three facilities: Verdi Hatchery and Washoe Rearing Ponds (Nevada Department of Wildlife) and Hagerman NFH (U.S. Fish and Wildlife Service). Federal participation in Pyramid Lake stocking was completely transferred from Hagerman to Lahontan NFH in 1975, which was also the last year of stocking by the state of Nevada. The David Dunn and Captain Dave Numana hatcheries, operated by the Pyramid Lake Paiute Tribe, began stocking in 1975 and 1982, respectively.

Stocking included releases made into both Pyramid Lake and the lower Truckee River below Derby Dam. Many different stocking procedures were used to increase survival of fish, including resting of distantly transferred fish, short-term (over several hours) tempering of fish, barging fish offshore, net-pen acclimation, night stocking, and stocking at different locations around the lake. However, most fish received no acclimation or tempering and were released from shore at two or three locations that were readily accessible to trucks.

Angling Regulations

In addition to the stocking variables such as species and strains of trout and numbers and sizes of fish released, changes in angling regulation affected the fishery. Between 1954 and 1987, angling regulations changed six times (Table 2). During the first 10 years, regulations were very liberal: the bag limit was five fish; there was no size limit, and bait use was allowed. Since 1964, special regulations have been in effect that restrict anglers to use of artificial lures and flies only, a daily bag limit of two or three fish, and a 381–483-mm minimum-size limit. These regulations were incorporated to allow stocked trout to reach larger sizes before they were harvested, thus taking advantage of the productivity of Pyramid Lake and its abundant forage base. The most noteworthy regulation change since 1964 has been the minimum-size limit, which has varied more than 100 mm.

TABLE 2.—Summary of fishing regulations at Pyramid Lake, Nevada, 1954–1987.

Time period	Bag limit	Minimum-size limit (mm)	Gear
Jan 1954–Dec 1963	5	None	Bait, lures, flies
Jan 1964–Dec 1966	3	483	Lures, flies
Jan 1967–Dec 1970	3	432	Lures, flies
Jan 1971–Dec 1976	2	483	Lures, flies
Jan 1977–Jun 1980	2	381	Lures, flies
Jul 1980–Jun 1982	2	457	Lures, flies
Jul 1982–present	2	483	Lures, flies

Creel Survey

Vital statistics for the trout sport fishery at Pyramid Lake have been collected through annual creel surveys (Table 3). Anglers were interviewed either by a roving census clerk while fishing or at a fixed check station as they left the lake. Data collected included hours fished; boat or shore fishing mode; composition, length, and weight of the catch; and recovery of tags. Most survey data were collected between October and April, coinciding with most of the annual angler use, although the season was open year-round. Between 1955 and 1978, surveys occurred on 25–100 d annually; census hours usually were between noon and sunset. Roving surveys were made on days of low angler use. Approximately 60% of the surveys were conducted on weekends.

Creel survey methods were standardized in 1978. Since then, all data have been collected at a fixed creel station by methods similar to those described by Robson (1960). The collection effort (hours) was stratified in a manner similar to the fishing pressure (Carlander and Di Dostanzo

TABLE 3.—Creel survey data for Pyramid Lake, Nevada, 1955–1987.

Fishing season	Creel survey days	Survey months	Number of anglers surveyed	Hours fished	Number of trout: Creeled	Number of trout: Returned to the lake
1955	7	Jan–May	797	3,166	171	
1955–1956	33	Oct–May	1,884	6,707	967	
1956–1957	74	Sep–Mar	4,999	22,348	2,108	
1957–1958	51	Sep–Feb	2,910	13,472	1,065	
1958–1959	70	Oct–Jun	7,080	36,421	4,633	
1959–1960	79	Oct–Mar	10,317	51,342	4,304	
1960–1961	70	Sep–Mar	7,428	35,023	2,367	
1961–1962	86	Sep–Mar	9,800	46,235	6,580	
1962–1963	57	Oct–Jun	3,884	18,842	1,564	
1963–1964	70	Sep–Apr	4,689	21,748	2,210	443
1964–1965	50	Oct–Mar	3,626	15,170	338	2,272
1965–1966	48	Oct–Apr	3,762	16,706	611	3,852
1966–1967	47	Oct–Apr	5,543	25,862	1,029	7,315
1967–1968	47	Oct–Apr	9,942	47,004	4,942	9,433
1968–1969	58	Oct–Apr	13,982	63,960	9,156	5,484
1969–1970	47	Sep–May	17,556	85,753	7,286	6,453
1970–1971	49	Oct–Apr	13,607	66,261	3,658	5,287
1971–1972	46	Oct–Apr	9,549	46,772	2,467	3,959
1972–1973	47	Oct–Apr	8,906	42,728	1,814	4,167
1973–1974	48	Oct–Apr	8,401	40,753	1,023	8,094
1974–1975	25	Oct–Apr	1,025	3,041	221	380
1975–1976	110	Oct–May	5,696	24,911	1,315	3,864
1977	42	Jan–May	5,186	27,842	1,062	1,243
1977–1978	61	Oct–Mar				
1978–1979	70	Oct–Apr	4,238	21,470	1,759	4,501
1979–1980	70	Oct–Apr	4,425	21,450	1,656	3,138
1980–1981	70	Oct–Apr	5,941	31,478	1,006	8,046
1981–1982	70	Oct–Apr	5,414	28,175	1,500	3,622
1982–1983	70	Oct–Apr	6,695	34,644	2,113	6,538
1983–1984	70	Oct–Apr	7,132	37,709	2,047	6,669
1984–1985	70	Oct–Apr	6,528	35,472	2,046	7,266
1985–1986	67	Oct–Apr	6,091	33,277	1,768	6,060
1986–1987	69	Oct–Apr	6,682	36,964	1,642	6,105

1958); four weekdays and six weekend days were randomly selected for each month. The fixed station was set up on the principal highway leading to Pyramid Lake from Reno. Survey hours were noon to sunset.

Angler use (angler-days) from 1954 until 1968 was determined by multiplying the number of seasonal tribal permits sold by the mean number of trips per permit, and this number was added to the daily permits sold. Since 1968, angler use was estimated from annual Nevada Department of Wildlife statewide angler-use surveys sent to 10–20% of all licensed anglers (McLelland and Burgoyne 1986). Harvest was estimated by multiplying the mean annual creel rate (fish/h) by the mean number of hours fished per day times the total annual angler-days.

The Fishery

Landing rates at Pyramid Lake were poorest between 1954 and 1964 (mean, 0.10 fish/h) and harvest ranged from 2,600 to 24,000 fish (mean, 12,000; Figure 2). During this period, annual stocking levels were only 45,000 to 175,000 fish over 125 mm, and most were rainbow trout. Annual angler returns of stocked fish to the creel (harvest divided by number of fish stocked) ranged from 2 to 5%.

Low landing rates and harvest occurred even though angling regulations were liberal (Table 2). Survival of rainbow trout was probably low due to the relative inability of this species to acclimate to the highly alkaline lake waters. This was documented by several bioassay and live-cage studies. Rainbow trout that were transported from distant hatcheries, particularly Hagerman NFH (10–12 h distance from the lake), did very poorly unless they were rested for several days before release. However, most releases from this facility were not rested; over 50% of the rainbow trout stocked originated from there.

Those rainbow trout that survived the initial stocking grew quite well, reaching approximately 300–400 mm after 1 year in the lake. Very few fish were caught in their subsequent years. Low returns after the first year were related to the preference of rainbow trout for shoreline areas, making them particularly vulnerable to anglers.

Annual mean length of angler-captured trout (rainbow and cutthroat trout combined) ranged from 350 to 410 mm; fish over 600 mm were rare (Figure 2). Cutthroat trout were less vulnerable to fishing and had substantial returns for 2–4 years following stocking. From 1954 to 1964, angler returns during the first year after release averaged 40% for cutthroat trout. However, only small numbers of this species were stocked during the 1950s and early 1960s, primarily because they were hard to rear and egg sources had not yet been developed.

Angler use from 1954 to 1964 ranged from 21,000 to 40,000 angler-days (average, 32,000; Figure 2). This low use was primarily related to poor seasonal landing rates of 0.05–0.14 fish/h. Also during this period, most captured fish were 300- to 400-mm rainbow trout. Many anglers apparently sought waters that offered better landing rates of similarly sized rainbow trout.

Based on the limited Lahontan cutthroat trout stocking that occurred in the 1950s, it was readily apparent that this species was more successful in Pyramid Lake than rainbow trout. Steps were taken to convert hatchery production to Lahontan cutthroat trout in the late 1950s by developing brood-stock populations in nearby lakes. Large-scale Lahontan cutthroat trout stocking essentially began in 1961, and rainbow trout stocking was gradually phased out by 1966 (Table 1). Landing rates improved substantially after 1964, and improvements can be primarily attributed to the shift from rainbow trout to Lahontan cutthroat trout stocking. The mean annual landing rate from 1964 to 1987 was 0.21 fish/h. Harvest also increased substantially after 1964 even under the more restrictive angling regulations. Angler use, influenced by angler success, was two to five times higher than pre-1964 levels.

The period from 1964 to 1971 was the most successful for the Pyramid Lake fishery. During this time, annual releases of 81,000 to 410,000 fish (6,700–14,000 kg) resulted in landing rates of 0.14–0.32 fish/h (Figure 2). Creel rates, lower due to minimum-size limits, ranged from 0.02 to 0.14 fish/h. Annual mean size of creeled trout averaged 472–592 mm, and fish over 600 mm accounted for 28% of the fish harvested. Harvest during the period started slowly but rose dramatically, reaching 84,000 fish in 1968. Annual angler returns averaged approximately 19% of the fish stocked when harvest was compared with releases 2 years previously.

Angler use also increased dramatically during this period, reaching 142,000 angler-days in 1970 (Figure 2). This occurrence was related to greatly improved landing rates and the large size of fish caught. Angler use was most closely tied to the size of fish caught. During the late 1960s, when creel and landing rates dropped, angler use continued to rise. However, the mean size of creeled fish and the catch of fish over 600 mm both increased from 1967 to 1970. Trophy catches (creeled fish over 600 mm) peaked at 55%, and the mean size of creeled fish was 599 mm in 1970. Apparently, anglers were enticed by the prospects of catching large Lahontan cutthroat trout. Increases in the human population size in western Nevada during the 1960s affected angler use to a lesser extent.

Landing and creel rates from 1972 to 1987 have vacillated unimpressively between 0.08 and 0.28 fish/h and 0.03 to 0.08 fish/h, respectively (Figure 2). Angler use dropped during the early 1970s with reductions in angler success. Since 1975, use has fluctuated between 58,000 and 96,000 angler-days. Closely associated with angler use and rate of success, harvest ranged from 13,500 to 29,000 fish for this period. Average annual angler return of creeled fish has ranged from 1 to 7% of the cutthroat trout released 2 years previously.

Although stocking rates increased several-fold in the 1970s and 1980s, landing rates have never reached the levels recorded in the 1960s. The tremendous success of the late 1960s may represent a phenomenon similar to that occurring in lakes when forage fish are introduced with a predator for the first time. Stocking of substantial numbers of Lahontan cutthroat trout began in 1961. Prior to that, mostly rainbow trout were stocked; these fish were harvested soon after release and inhabited mainly the littoral regions of the lake, feeding primarily on invertebrates. It appears the population of tui chub flourished in the 20–30 years that a large pelagic predator was absent. It was not until cutthroat trout were reintroduced that this tremendous food base was tapped. Survival rates were also high for these initial plants because there were few large cutthroat trout to prey on the small stocked fish.

The strains of Lahontan cutthroat trout changed concurrently with declines in the sport fishery. During the 1960s, the Heenan Lake strain was the predominant fish stocked. This cutthroat trout, although phenotypically resembling the Lahontan cutthroat trout, is slightly hybridized with rainbow trout (Gall and Loudenslager 1981). Cutthroat trout stocked after 1971 were mostly the Summit Lake strain. This genetically pure Lahontan cutthroat trout apparently has been isolated in a small freshwater lake environment in the absence of forage fish for several thousand years (Behnke and Zarn 1976). Comparisons between these two strains are available for 1958, 1969, and 1981, when tagged groups with similar rearing and stocking histories were released. Only from the 1969 releases were angler returns substantially higher for the Heenan Lake strain. Both strains had similar returns from the 1958 and 1981 release groups. Thus, based on the limited tagging data, strain cannot be singled out as a major factor leading to the low return rates of the 1970s and 1980s.

Angling regulation changes at Pyramid Lake have had a profound effect on the trout fishery. Under liberal regulations in the 1950s and early 1960s (Table 2), most rainbow trout were caught within 1 year of their release; mean annual size of creeled trout averaged less than 425 mm (Figure 2). Since 1964, however, special regulations have been in effect, restricting gear use, bag limit, and minimum size. These regulations were not compatible with the rainbow trout fishery because very few fish of this species survived to reach the minimum size limit. The Lahontan cutthroat trout fishery benefited from the special regulations; these fish were afforded protection and grew rapidly as they fed on the abundant forage base in the lake. Average annual size of creeled fish increased dramatically under special regulations, ranging from 475 to 600 mm (Figure 2).

The most noteworthy regulation changes after 1964 were in minimum-size limits, which varied from 381 to 483 mm (Table 2). Creel rates always have increased when the minimum-size limit was reduced and have decreased when the size limit increased. However, results were not consistent for all years under a specific size-limit regulation. Total catch rate and catch of trophy fish do not appear to have been affected by minimum-size-limit changes.

Low angler returns and landing rates are disturbing because the lake has the potential to produce many more cutthroat trout. The trout harvest since 1964 has averaged only 40,000 kg (27,600 fish) annually, whereas Galat (1983) estimated the potential trout production in Pyramid Lake to be 265,000–3.5 million kg annually. Behnke (unpublished) estimated the combined production for Winnemucca and Pyramid lakes in the late nineteenth century at 454,000 kg/year. The low harvest and landing rates suggest that survival of stocked Lahontan cutthroat trout has been low. It appears that these fish, strains from various sources, are not well suited for the unique environmental conditions in Pyramid Lake. Attempts to improve angler returns by use of various stocking protocols, including release size, stocking time, stocking date, release location, and acclimation, have generally been unsuccessful. Some techniques, like larger release size, result in higher returns; however, even the best release groups since the mid-1970s have generated angler return rates of less than 5%.

Future improvement of fishing at Pyramid Lake seems to depend on development of a Pyramid Lake strain. Work towards achievement of this goal has been underway since 1981, when the Pyramid Lake Paiute Tribe first took eggs from fish surviving in the lake. Each year since then, an increasing percentage of the annual production has come from eggs taken from cutthroat trout surviving in the lake; this figure reached 100% in 1987. There is still interest in introducing Lahontan cutthroat trout from Donner Creek, Utah. These fish are believed to be related to the original Pyramid Lake strain (Hickman and Behnke 1979); however attempts to establish a brood stock from this small population have been unsuccessful.

The restored Pyramid Lake cutthroat trout fishery is maintained solely through hatchery production. If the fishery is ever to achieve a semblance of the production potential, natural recruitment of juveniles from the Truckee River will be necessary (Sigler and Kennedy 1978). Galat (1983), using a model developed by Innis et al. (1981), estimated that, under ideal conditions, the Truckee River could produce 4 million fingerlings. Conditions are far from ideal, however, and large-scale habitat rehabilitation would be necessary to achieve this goal. Conditions in the Truckee River presently favor natural recruitment from a fall spawning run. A small fall run has been observed over the last several years at Pyramid Lake and work to expand this run is in progress.

Acknowledgments

Funding and technical support came from the U.S. Fish and Wildlife Service, National Fishery Research Center–Seattle and Great Basin Complex. Many of the data for this study were collected by personnel from Pyramid Lake Fisheries, the resource management branch of the Pyramid

Lake Paiute Tribe. Creel survey data prior to 1975 were collected by personnel from the Nevada Department of Wildlife.

References

Behnke, R. J., and M. Zarn. 1976. Biology and management of threatened and endangered western trout. U.S. Forest Service General Technical Report RM-28.

Benson, L. V. 1978. Fluctuations in the level of pluvial Lake Lahontan for the past 40,000 years. Quaternary Research (New York) 9:300–318.

Carlander, K. D., and C. J. Di Costanzo. 1958. Sampling problems in creel census. Progressive Fish-Culturist 20:73–81.

Cole, G. A. 1979. Textbook of limnology. Mosby, St. Louis, Missouri.

Galat, D. L. 1983. Primary production as a predictor of potential fish production: application to Pyramid Lake, Nevada. Doctoral dissertation. Colorado State University, Fort Collins.

Galat, D. L., E. L. Lider, S. Vigg, and S. R. Robertson. 1981. Limnology of a large, deep North American terminal lake, Pyramid Lake, Nevada. Hydrobiologia 82:281–317.

Gall, G. A., and E. J. Loudenslager. 1981. Biochemical genetics and systematics of Nevada trout populations. Report to Nevada Department of Wildlife, Reno.

Hickman, T. J., and R. J. Behnke. 1979. Probable discovery of the original Pyramid Lake cutthroat trout. Progressive Fish-Culturist 41:135–137.

Innis, G. S., D. F. Hanson, and J. W. Haefner. 1981. A simulation model of management alternatives in a freshwater fishery. Ecological Modeling 12:267–280.

Knack, M. C., and O. C. Stewart. 1984. As long as the river shall run. University of California Press, Berkeley.

Knoll, J., D. L. Koch, R. Knoll, J. Sommer, L. Hoffman, and S. Lintz. 1979. Physiological adaptations of salmonid fishes (*Salmo clarki henshawi*, *Salmo gairdneri*, and *Oncorhynchus kisutch*) to alkaline saline waters and their toxic effects. Desert Research Institute, Bioresources Center Publication 50009, Reno, Nevada.

La Rivers, I. 1962. Fishes and fisheries of Nevada. Nevada State Fish and Game Commission, Carson City.

McLelland, L., and M. Burgoyne. 1986. Statewide angler use and harvest survey report. Nevada Department of Wildlife, Reno, Nevada.

Robson, D. S. 1960. An unbiased sampling and estimating procedure for creel census of fishermen. Biometrics 16:261–277.

Scoppettone, G. G., M. Coleman, and G. A. Wedemeyer. 1986. Life history and status of the endangered cui-ui of Pyramid Lake, Nevada. U.S. Fish and Wildlife Service, Fish and Wildlife Research 1.

Sigler, W. F., and J. L. Kennedy, editors. 1978. Pyramid Lake ecological study. W. F. Sigler and Associates, Logan, Utah.

Snyder, J. D. 1917. The fishes of the Lahontan system of Nevada and northeastern California. U.S. Bureau of Fisheries Bulletin 35:31–85.

Sumner, F. H. 1940. The decline of Pyramid Lake fishery. Transactions of the American Fishery Society 69:216–224.

Townley, J. M. 1980. The Truckee basin fishery, 1844–1944. Desert Research Institute, Water Resources Center Publication 43008, Reno, Nevada.

Trelease, T. J. 1969a. The death of a lake. Nevada Outdoors and Wildlife Review 3(1):4–9.

Trelease, T. J. 1969b. The rebirth of a lake. Nevada Outdoors and Wildlife Review 3(1):10–14.

USFWS (U.S. Fish and Wildlife Service). 1983. Restoration of a reproductive population of Lahontan cutthroat trout (*Salmo clarki henshawi*) to the Truckee River/Pyramid Lake system. USFWS, Reno, Nevada.

Vigg, S. 1981. Species composition and relative abundance of adult fish in Pyramid Lake, Nevada. Great Basin Naturalist 41:395–409.

Wetzel, R. G. 1978. Limnology. Saunders, Philadelphia, Pennsylvania.

Wheeler, S. S. 1974. The desert lake: the story of Nevada's Pyramid Lake. Caxton Printers, Caldwell, Idaho.

American Fisheries Society Symposium 4:116–120, 1988

Status of a Hybridized Population of Alvord Cutthroat Trout from Virgin Creek, Nevada

DENNIS TOL

U.S. Bureau of Land Management, 705 East 4th Street, Winnemucca, Nevada 89445, USA

JIM FRENCH

Nevada Department of Wildlife, 170 East Nimitz, Winnemucca, Nevada 89445, USA

Abstract.—A project was developed and conducted in 1986 to recover and protect what was thought to be a relict population of Alvord cutthroat trout, an undescribed subspecies of *Salmo clarki*, from Virgin Creek in the Alvord Basin of northwest Nevada and southeast Oregon. It was determined through electrophoretic analysis that the population was thoroughly hybridized with rainbow trout *Salmo gairdneri*. Electrophoresis also revealed that the trout from Virgin Creek, although destinctive in size and coloration, did not differ genetically from Lahontan cutthroat trout *Salmo clarki henshawi* at five diagnostic loci. Previous meristic analysis had also indicated similarities with *S. c. henshawi*. It is highly likely, considering both the topography and past geological events, that the Virgin Creek cutthroat trout originated from Mahogany Creek and Summit Lake, which were once part of the Lahontan drainage basin. Twenty-six individuals that displayed general morphometric characters of cutthroat trout were selected and transplanted into Jackson Creek, Nevada.

The Alvord Basin in northeastern Nevada and southeastern Oregon was thought to have contained a separate subspecies of cutthroat trout *Salmo clarki*, based on the geographic isolation of the fish and the meristic analysis of specimens from early sampling efforts (Behnke 1979; Figure 1). However, recent scientific inventories and surveys of streams within the Alvord Basin have failed to produce cutthroat that have not hybridized with rainbow trout *Salmo gairdneri* or that are dissimilar in genetic, morphometric, and meristic characters to the Lahontan cutthroat trout *Salmo clarki henshawi* of the adjacent Lahontan drainage basin.

Rumors of a unique trout in the upper watershed of Virgin Creek, Nevada, prompted a 1983 survey of that stream by the Nevada Department of Wildlife (NDW). Taxonomic and genetic evaluations of specimens collected randomly in a 1984 survey indicated that the population had hybridized with rainbow trout and had morphometric characters that differed from fluvial populations of Lahontan cutthroat trout (e.g., in coloration and size). In 1985, the population was again sampled for genetic evaluation, with the intent of selecting Alvord cutthroat trout that might be pure, based on the morphometric characters of jaw length, body coloration, and spot size, shape, and pattern. No genetically pure fish were discovered. Three F_1 backcrosses and a rare *Pgk-2(55)* allele were found in the 1985 sample (NDW, unpublished data).

Three characters distinguish the Virgin Creek population from other known fluvial cutthroat trout populations in the Alvord Basin, the Coyote Basin, and the Humboldt drainage in the Lahontan Basin. First, Virgin Creek cutthroat trout are long-lived, attaining ages of 5–7 years (Behnke 1986). Second, they are large, reaching total lengths of up to 63 cm. Third, spawning males acquire a broad rose- or brick-colored band from opercle to caudal peduncle.

Behnke (1979) provided the following meristic data for specimens collected from lower Virgin Creek by Carl Hubbs, in 1934, and by others:

(1) there were less than 50 relatively large round spots concentrated posteriously above the lateral line;
(2) few spots were on the caudal fin;
(3) gillraker numbers averaged 23.4;
(4) basebranchial teeth were poorly developed or absent;
(5) lateral series scales averaged 135;
(6) pyloric caeca averaged 42;
(7) vertebrae averaged 61.2.

Hubbs, according to field notes about his 1934 collection, observed specimens with white leading edges on the pelvic and anal fins and an orange-colored tip on the anterior dorsal fin.

Although cutthroat trout were reported to exist in the Alvord Basin waters of Trout, Cottonwood, Thousand, and Virgin creeks (Hubbs, 1934, un-

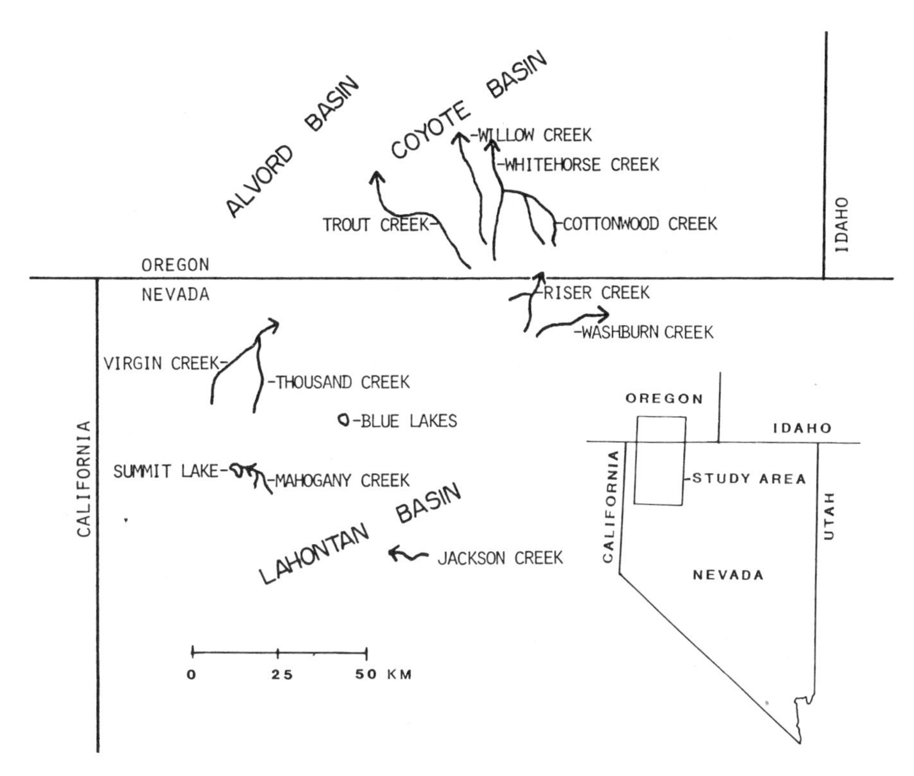

FIGURE 1.—Location of streams, lakes, and basins within the study area.

published) and Blue Lakes as late as 1921 (NDW, unpublished data), these populations no longer exist (Figure 1). Behnke (Colorado State University, unpublished) identified three specimens from Virgin Creek as pure Alvord cutthroat trout based on meristic similarities with Hubbs' 1934 fish, which Behnke assumed were not hybridized.

The current study was developed and conducted in 1986 to determine if genetically pure individuals still existed in Virgin Creek and to transplant either genetically pure individuals, if they existed, or individuals selected for their cutthroat trout appearance (based on morphological characters) from Virgin Creek to Jackson Creek, Nevada, a barren stream in the Lahontan Basin.

Study Area

Virgin Creek is a small spring-fed creek that originates at an elevation of 1,743 m and is approximately 32.2 km long. It has a relatively low mean gradient (1.5%; NDW Stream Survey) and is perennial only in those reaches below spring heads. The aquatic habitat has been influenced greatly by beaver dams. There are also deep holes or tanks along the channel that are fed by subsurface flows. These tanks and the nearby channels can be as much as 3.3 m deep and filled to the top with water. Aquatic vegetation is abundant. The channel averages 1.2 m wide and 0.14 m deep. There is a rock slide midway in the stream course that diverts the flow underground. This slide acts as an effective barrier to upstream and downstream movement and creates two separate populations of trout in Virgin Creek. A common element in the historical memory of the local public is the large red trout that existed in the upper half of Virgin Creek above the rock slide.

The stocking record for Virgin Creek is poorly documented. The first scientific collections were made by Hubbs in the fall of 1934. His field notes documented a statement by Tom Duffurena, a local rancher, indicating that no stocking of rainbow trout had taken place in Virgin Creek until he planted 6,000 fingerling rainbow trout in the lower reaches during the fall of 1933. However, Hubbs found cutthroat–rainbow trout hybrids and rainbow trout as large as 29 cm in 1934, 1 year later. He attributed this to a high growth rate and stocking of hybrids in Nevada.

The current owner of Virgin Creek Ranch, H. W. Wilson, stated that in 1942 his father planted rainbow trout into Alkali Reservoir, on the headwaters of Virgin Creek (personal communication). These rainbow trout spilled over and populated the upper half of Virgin Creek, thereby establishing rainbow trout throughout the system. Wilson also mentioned the possibility that cutthroat trout were stocked into the lower portion of Virgin Creek after a period of severe drought. No mention was ever made of cutthroat trout having been stocked in Virgin Creek above the rock slide.

Jackson Creek is a perennial spring creek which is 13.5 km long and has a much steeper gradient (9%) than Virgin Creek. Its average width is approximately 2.4 m, and its average depth is approximately 0.09 m in the fall. The stream bottom is predominately rubble with approximately 28% spawning gravels (U.S. Bureau of Land Management stream survey). The pools are relatively shallow, and there is no slow-moving water or aquatic vegetation. It does not provide habits like those of Virgin Creek, and it is barren of fishes. Food organisms available to fish in Jackson Creek are predominately benthic and terrestrial insects.

Methods

Two attempts were made in 1986 to collect fish from Virgin Creek for transplant and for electrophoretic analysis. The first transplant attempt, in August, was canceled because of excessive deaths caused by high water temperatures and low dissolved oxygen. The dead fish were sent to University of California (UC) in Davis for electrophoretic analysis.

The second attempt, in October, was successful, 60 fish were collected from Virgin Creek. Based on the similarity of their appearance to cutthroat trout, 26 of these fish were selected for transplant into Jackson Creek, Nevada. The selection was based on obvious and easily evaluated external characteristics, such as spotting pattern, spot size and shape, jaw length, and body coloration, all of which were presumed to be indicative of the native Alvord cutthroat trout. Twenty-five fish were successfully transplanted into Jackson Creek on October 3, 1986.

Photographs and field notes were taken of 13 of the transplanted fish for later correlation with electrophoretic data. For the electrophoretic analysis, a small piece of muscle tissue was taken from each fish just below the dorsal fin; it was frozen on dry ice and transported to the Department of Animal Science, UC–Davis, where electrophoretic analysis was done under contract with NDW. Electrophoresis followed the procedures of Busack et al. (1979) and was used to determine a genetic profile for each individual at five diagnostic loci (*Ck-2*, *Gl-1*, *Lgg-1*, *Mdh-2*, and *Pgk-2*). This profile was compared to standard genetic profiles for cutthroat and rainbow trout (E. J. Loudenslager and G. A. E. Gall, University of California, unpublished data).

TABLE 1.—The average proportion of rainbow trout and Lahontan cutthroat trout alleles from four samples taken from the Virgin Creek population (sample sizes in parentheses).

	Sample date			
Species	1984 (38)	1985 (5)	Aug 1986 (25)	Oct 1986 (36)
Rainbow trout	0.97	0.48	0.50	0.57
Cutthroat trout	0.03	0.52	0.50	0.43

Results and Discussion

No pure cutthroat trout were found during the recovery effort. Electrophoretic analysis of all fish sampled showed a history of hybridization between cutthroat and rainbow trouts (Table 1).

The 1984 collection from Virgin Creek was made without regard to the appearance of the fish. The population was sampled once in 1985 and twice in 1986 with the intent of selecting fish for their cutthroat trout appearance in hopes of discovering genetically pure Alvord cutthroat trout. The individual fish from the October 1986 sample whose appearance was closest to the "ideal" cutthroat trout phenotype had no cutthroat trout alleles at the diagnostic loci. Devin Bartly (personnel communication), who conducted the electrophoretic analysis under the supervision of G. Gall, indicated that hybridization had progressed to such a degree that phenotype was no longer an indicator of the fish's genotype. The presence of rainbow trout alleles at the diagnostic loci indicated that rainbow trout alleles probably existed at other unexamined loci whether or not they affected morphology.

The Jackson Creek population will be examined for evidence of successful spawning. Follow-up evaluations should show if this new population breeds true phenotypically or if it expresses a broad spectrum of morphological and meristic characters that range between those of cutthroat trout and rainbow trout. It may also be possible to determine if size, life span, and possibly some of the coloration are environmentally influenced.

Conclusions

For all practical purposes, the Alvord cutthroat trout, in its genetically pure state, can be considered extinct, unless there remains an undiscovered population in a stream somewhere else in the Alvord Basin. This project shed little light on the original meristic and electrophoretic characteristics of Alvord cutthroat trout, and the existence of a separate subspecies in the Alvord Basin may be questioned. The genetic identity of the Virgin Creek trout remains in doubt. The fish appear to represent an introgressed population or "hybrid swarm" derived from rainbow trout and a Lahontan-like cutthroat trout.

Many questions generated by past speculation about the Alvord cutthroat trout were left unanswered. It is highly likely that the Virgin Creek cutthroat trout came from the Lahontan drainage by way of Mahogany Creek to the southwest. A great landslide blocked off the Mahogany Creek watershed from the Lahontan basin and created Summit Lake. It appears that Lahontan cutthroat trout were trapped in Summit Lake and Mahogany Creek at that time. During wet years, Summit Lake has the potential to overflow into the Alvord Basin through Virgin Creek. There is no record of this occurring in recent times, but Summit Lake came very close to spilling over in 1983.

Inconsistencies in Hubbs' field notes seem to indicate that there may have been a hybridized population of cutthroat trout in lower Virgin Creek prior to the first recorded plant of rainbow trout. It is unlikely that rainbow trout moved upstream through the rockslide, but it is always possible that they were stocked above the rockslide. Our project found such thorough hybridization above the slide that hybridization may well have taken place before rainbow trout moved into Virgin Creek from Alkali Reservoir after 1942 (Gall, personal communication).

The Virgin Creek fish display white leading edges on their anal and pelvic fins and orange tips on the anterior of their dorsal fins. These characteristics, along with the brick-red coloration, are considered characteristics of inland ("redband") rainbow trout, although there is no evidence that such fish ever existed naturally in the Alvord Basin (W. Hosford, Oregon Department Fish and Wildlife, personal communication).

It is possible that the bright coloration, longevity, and large size of these fish may be chiefly due to the environment of Virgin Creek. However, longevity could also derive from a genetic relationship to the probable parental stock from Summit Lake, which reaches 7–8 years of age. Space in Virgin Creek is limiting, as it is in other small steams of the Great Basin; however, food is abundant. Virgin Creek, in places, is dammed by beavers and is extremely slow moving, with holes as deep as 3.3 m. Crustaceans (Branchiopoda, Copepoda, and Malacostraca) are abundant and presumed to make up a large part of trout diets. The flesh color of the Virgin Creek fish is bright orange. Size and external coloration could be affected by the diet.

Behnke (1979) stated that average gillraker numbers indicate that the Virgin Creek trout were derived from Lahontan cutthroat trout of the Lahontan Basin. Other meristic and morphological characters, however, such as spotting pattern, poor development of basibranchial teeth, and low scale counts, suggest a long separation from Lahontan stocks, as well as hybridization with rainbow trout.

A Lahontan cutthroat trout ancestral link is certainly plausible, and highly likely, considering the topography and past geological events in the vicinity of the headwaters of Virgin Creek. Cutthroat trout from Washburn Creek in the Quinn River drainage of the Lahontan Basin have spot shape and pattern, body conformation, and coloration similar to those described for the Alvord cutthroat trout. The cutthroat trout from Washburn Creek were tested electrophoretically in 1984 and were determined to be pure Lahontan cutthroat trout (NDW, unpublished data). The rare *Pgk-2(55)* allele, which was found in the Virgin Creek trout, was also found in pure Lahontan cutthroat trout from Riser Creek in the Quinn River drainage.

Willow, Whitehorse, and Little Whitehorse creeks in the Coyote Basin, which is adjacent to, and may have been connected to, the Alvord Basin, contain pure Lahontan cutthroat trout (NDW, unpublished data). All other trout populations sampled in the Alvord Basin are either hybridized or show no cutthroat trout influence. The cutthroat trout in Willow and Whitehorse creeks may have been transplanted from the Quinn or Humboldt river drainages in the Lahontan Basin. The Virgin Creek trout may have originated from Lake Lahontan parental stock. Behnke (1979) speculated that Alvord cutthroat trout moved from the Alvord Basin to the Willow–Whitehorse Creek system at the headwaters of the two drainages and that the two groups subsequently diverged. Recent work on the genetics of these fish, however, indicates it is more likely that the Willow–Whitehorse Creek fish resulted from an unrecorded transplant of Lahontan cutthroat trout from the Humboldt or Quinn river drainages. Behnke's (1979) meristic analysis of Willow Creek, White Horse Creek, and Humboldt Basin populations suggests a close relationship among these fish, although Behnke did not propose one.

Our study indicates that isolated trout populations from closed desert drainage basins, like the Alvord Basin, must be evaluated with caution. Catastrophic events capable of wiping out trout populations occur with regularity along small desert streams. Many of these streams that flow into desert valleys or sumps cannot be naturally repopulated. Unregulated stocking and transplanting of fish, which was widespread during the early 1900s, established populations in many barren streams. It is quite possible that the Alvord cutthroat trout was extinct before Europeans arrived in northern Nevada; yet recollections and incomplete records indicate, tantalizingly, that a cutthroat trout did exist in the Alvord Basin. It is more likely that genetically pure populations were extirpated before we became aware of their potentially unique status.

Acknowledgments

This project was coordinated by the Nevada Department of Wildlife with the cooperation of the U.S. Bureau of Land Management and the U.S. Fish and Wildlife Service, Sheldon Antelope Refuge. The electrophoretic analyses were done under contract with the Department of Animal Science, University of California, Davis.

References

Behnke, R. J. 1979. The native trouts of the genus *Salmo* of western North America. Report to U.S. Fish and Wildlife Service, Denver, Colorado.

Behnke, R. J. 1986. Alvord cutthroat trout. Trout 27(2): 50–54.

Busack, C. A., R. Halliburton, and G. A. S. Gall. 1979. Electrophoretic variation and differentiation in four strains of domesticated rainbow trout. Canadian Journal of Genetics and Cytology 21:81–94.

Loudenslager, E. J., and G. A. E. Gall. 1980. Geographic patterns of protein variation and subspeciation in cutthroat trout, *Salmo clarki*. Systematic Zoology 29:27–42.

American Fisheries Society Symposium 4:121–127, 1988

Bonneville Cutthroat Trout: Current Status and Management

DONALD A. DUFF

U.S. Forest Service, 324 25th Street, Ogden, Utah 84401, USA

Abstract.—Only one trout subspecies, the Bonneville cutthroat trout *Salmo clarki utah*, is endemic to the Bonneville Basin, the largest endorheic basin in the Great Basin of western North America. The subspecies was historically abundant throughout all suitable habitat of a vast area of Utah, Wyoming, Nevada, and Idaho until about 8,000 years ago, when the desiccation of ancient Lake Bonneville occurred. During this century, the subspecies suffered a catastrophic decline from introductions of nonnative trouts and habitat alteration. Today, only 41 populations, primarily limited to small headwater streams, are known to be genetically pure. Habitat conditions existing in waters containing the Bonneville cutthroat trout are marginal, and the streams are generally small with depleted flows. Yet, in many of these streams, Bonneville cutthroat trout still survive. Introductions of nonnative trout have been largely unsuccessful in these marginal streams, but they have led to the demise of the native cutthroat trout. The Bonneville cutthroat trout has promising possibilities for enhancing wild trout fishery management programs within the Great Basin states. As a result of the subspecies' limited distribution and present threats to its survival, protection of the few remaining populations is being considered. This paper summarizes the current status and management direction for this subspecies.

The Bonneville Basin occupies an area of about 14 million hectares in the Great Basin of the western USA, primarily in central and northern Utah, but also in eastern Nevada, southeastern Idaho, and southwestern Wyoming. The ancient Pleistocene Lake Bonneville, for which the basin was named, occupied 5 million hectares to a maximum depth of about 333 m. At this size, the lake's total volume of water was equivalent to the normal outflow to the oceans of all rivers of the USA for a period of 6 years. The lake eventually spilled north into the Snake River drainage through Red Rock Pass in southeastern Idaho. The Great Salt Lake is a remnant of Lake Bonneville. Major river drainages within the basin today are those of the Bear, Ogden, Weber, Provo, and Sevier rivers. It was in the aquatic habitat of this basin that the native Bonneville cutthroat trout *Salmo clarki utah* developed its unique form.

The Bonneville cutthroat trout is endemic to the Bonneville Basin and has suffered a catastrophic decline in the 20th century as a result of human land-management activities. At one time, this fish was presumed extinct; however, recent efforts by several individuals and agencies have provided new information on the existence of a few relict populations, and new hope has been generated for the future of the Bonneville cutthroat trout (Hickman 1978).

The fish fauna of the Bonneville Basin, although relatively depauperate, is the most extensive of the interior Great Basin drainages. There are one genus, seven species, and one subspecies (*S. c. utah*) of fish endemic to the Bonneville Basin. Limited endemism exhibited by Bonneville fishes suggests that mixing of adjacent faunas occurred during Pliocene and Pleistocene times. The most obvious of these faunal connections was with the upper Snake River above Shoshone Falls. The Bonneville cutthroat trout was probably derived from the cutthroat trout of the upper Snake River that became isolated by Shoshone Falls from the cutthroat trout of the lower Snake and upper Columbia Rivers (Hickman 1978).

The Bonneville cutthroat trout was of great importance to Indians in Utah as a source of food and to early European settlers, both for sustenance and commerce. The former abundance of this trout has been vividly documented (Hickman 1978). Probably the most detrimental modern factor causing the rapid decline of the Bonneville cutthroat trout since the civilizing impact of humans in the Bonneville Basin has been indiscriminate introductions (stockings) by wildlife agencies of nonnative trouts, particularly the rainbow trout *Salmo gairdneri*. Virtually every stream in the Bonneville Basin capable of supporting trout has been stocked with nonnative trouts. Hybridization of the Bonneville cutthroat trout with rainbow trout and other interior subspecies of cutthroat trout has led to almost complete elimination of pure populations of this subspecies.

Another significant impact on the survival of the Bonneville cutthroat trout is from physical

TABLE 1.—Characters used to distinguish pure populations of Bonneville cutthroat trout (Hickman 1978).

Character	Range	Mean	Description
Scale counts			
Lateral series	133–183	160	
Above lateral line	33–46	38	
Gill rakers	16–24	19	
Pyloric caeca	25–54	35	
Basibranchial teeth	1–50	[a]	
Coloration			One of the more somber hued cutthroat trout, the Bonneville subspecies does not develop the brilliant colors of some others. Typically has orange slash marks at throat.
Spotting			Large, sparse spots evenly distributed over the body. Bonneville cutthroat trout of the Snake Valley region tend to have more spots.

[a]Present in at least 90% of the population.

alterations of aquatic–riparian habitat by humans. The Bonneville cutthroat trout has been unable to readily adapt to these modifications. The most detrimental effects have been through elimination of streambank riparian vegetation, streambank trampling and erosion, sedimentation of streambed gravels, and loss of quality pool–riffle characteristics. These impacts have been caused primarily by livestock grazing, herbicide spraying, mining activities, road construction, and irrigation diversion. Subsequent alternating climatic conditions (floods and droughts) have accelerated the decline and loss of usable aquatic–riparian habitats in watersheds once used by this subspecies.

Taxonomy

Early ichthyologists named many species of cutthroat trout based on local varieties and did not recognize the range of morphological variability within a single species. The Bonneville cutthroat trout was so named as a species. Native trout of Utah Lake were mentioned in the report of the 1776 Dominguez–Escalante Expedition, and trout of the Bear River drainage were noted by Townsend in 1833 (Thwaits 1907). Suckley (1874) proposed *Salmo utah* to distinguish the trout of Utah Lake from other trout of the Bonneville basin called *S. virginalis*. This is the earliest reference to *S. utah* applied solely to trout of the Bonneville Basin (Hickman 1978).

Several states have been involved in enzyme electrophoresis to assist in identification of Bonneville cutthroat trout populations (Thompson 1987). Martin (1982) used sorbitol dehydrogenase (SDH) to separate different cutthroat trout enzyme stains, and Loudenslager and Gall (1980) used biochemical systematics to separate different forms of the Bonneville cutthroat trout. Behnke (1976, 1979, 1980), Hickman (1978), and May et al. (1978) have previously summarized taxonomic and biological information on this subspecies.

Although meristic and electrophoretic analyses have been used, researchers concluded that none of the taxonomic characteristics or techniques presently known can be used to positively identify an individual or a sample as genetically pure Bonneville cutthroat trout. An overlap in taxonomic characters exists due to the lack of long and complete isolation needed to evolve unique genetic differences in the various subspecies. However, there are well-defined average differences that separate the Bonneville cutthroat trout from nonnative trout that have been previously introduced into the Bonneville basin and could hybridize with the native trout (Table 1). No single character is significantly different from values for the other cutthroat trout subspecies, but, if taken collectively, these characteristics will usually distinguish the Bonneville cutthroat trout from other cutthroat trout subspecies (Hickman 1978). According to taxonomists, these characteristics most likely represent the original strain of the Bonneville cutthroat trout.

Using taxonomic and stock-purity analyses, Binns (1977, 1981) graded Bonneville cutthroat trout stocks in Wyoming with letter symbols to indicate stock purity (Table 2). While purity evaluations cannot be assessed with complete accuracy, they can be used to judge "pure" graded stocks when compared to original Bonneville cutthroat trout characteristics (Binns 1981).

Status

Today only 41 populations of pure Bonneville cutthroat trout are recognized (Figure 1). Although 39 populations are limited to headwater streams, there are two lake populations. The Bonneville cutthroat trout is not abundant in most

TABLE 2.—Purity ratings used in Wyoming to classify the Bear River strain of the Bonneville cutthroat trout (Binns 1981).

Purity rating	Characteristic
A	Pure stock (stock has the highest probability of representing the original Bonneville cutthroat)
B	Essentially pure stock
C	Stock is good representative of the Bonneville cutthroat, but some hybridization is evident
D	Stock is fair representative, but with definite hybridization evident
E	Stock has not been examined by taxonomist
F	Stock is poor representation; obvious hybrid characters present

of the streams because of generally poor aquatic–riparian habitats. In 1978, only 14 streams were known to contain the subspecies in 70 km of stream within the Bonneville basin. Since then, through interagency surveys, additional populations have been found in isolated streams. Other streams received transplanted stocks to expand the range of the subspecies. Today, the 39 stream populations in four states occur in about 302 km of stream. Several streams received transplanted stocks in the mid-1970s, but habitats were marginal, and fish populations were eliminated during the 1977 drought and spring floods of 1983–1984. Many agencies, organizations, and individuals have recognized the precarious status of this subspecies (e.g., Miller 1972; Behnke 1973; Holden et al. 1974). The U.S. Fish and Wildlife Service considers Bonneville cutthroat trout a candidate taxon for "threatened" status.

Management

State wildlife agencies and federal land management agencies have established a variety of cooperative programs for the preservation of the Bonneville cutthroat trout. These programs have expanded stocks within the Bonneville Basin.

The subspecies evolved in ancient Lake Bonneville, suggesting a genetic adaptation to lacustrine environments. Additionally, historical accounts often noted the large size of cutthroat trout in the Bonneville Basin. Although most remaining populations are limited to headwater streams, this subspecies may retain genetic characteristics that produce large fish when given the appropriate habitat. Management efforts to test this hypothesis are being implemented in Utah and Wyoming, with emphasis on lacustrine habitats. Lake Alice, Wyoming, and Bear Lake, Utah (Nielson and Lentsch 1988, this volume), represent lakes with self-sustaining stocks of Bonneville cutthroat trout.

The Bonneville cutthroat trout has been found in small headwater streams with degraded habitat and warm summer water temperature (21°C). Growth can be considered good (lengths of 320–350 mm at age 4 are common), although habitat conditions are marginal (eroding stream banks, warm water temperatures, low stream flows). Under similar conditions, nonnative trouts such as rainbow or brook trout *Salvelinus fontinalis* often stunt and are otherwise unacceptable to anglers (Hickman 1978).

Fishing for a trout as apparently scarce as Bonneville cutthroat trout may be questioned, but Behnke and Zarn (1976) have suggested that no rare or endangered trout has become so because of overfishing. Habitat conditions and introductions of nonnative trouts have apparently been responsible for declines.

Because populations of Bonneville cutthroat trout are limited, special angling regulations may offer the opportunity for angler use. Special regulations include slot limits, catch and release (no kill), creel limits, or other innovative strategies. These regulations allow angler use but provide protection while stocks continue to increase. Complete closure of streams to angling would also provide protection to diminished stocks, but this extreme restriction often allienates a portion of potential program support. Some states implemented several of these strategies in the early 1980s.

A summary of existing Bonneville cutthroat trout populations and current management efforts, by state, follows.

Idaho

Three pure populations in tributaries to the Bear River in the Caribou National Forest are being cooperatively managed by the Idaho Department of Fish and Game and the U.S. Forest Service (USFS). Livestock grazing systems are being revised to improve stream and riparian habitat condition. Bear Lake, which spans the Idaho–Utah boarder, also contains the only lake population of the Bonneville cutthroat trout in Idaho. Angling is allowed.

Wyoming

Seven pure populations in tributaries to the Bear River in the Bridger–Teton National Forest and the U.S. Bureau of Land Management (BLM) Kemmerer Resource Area are being managed cooperatively by the Wyoming Game and Fish

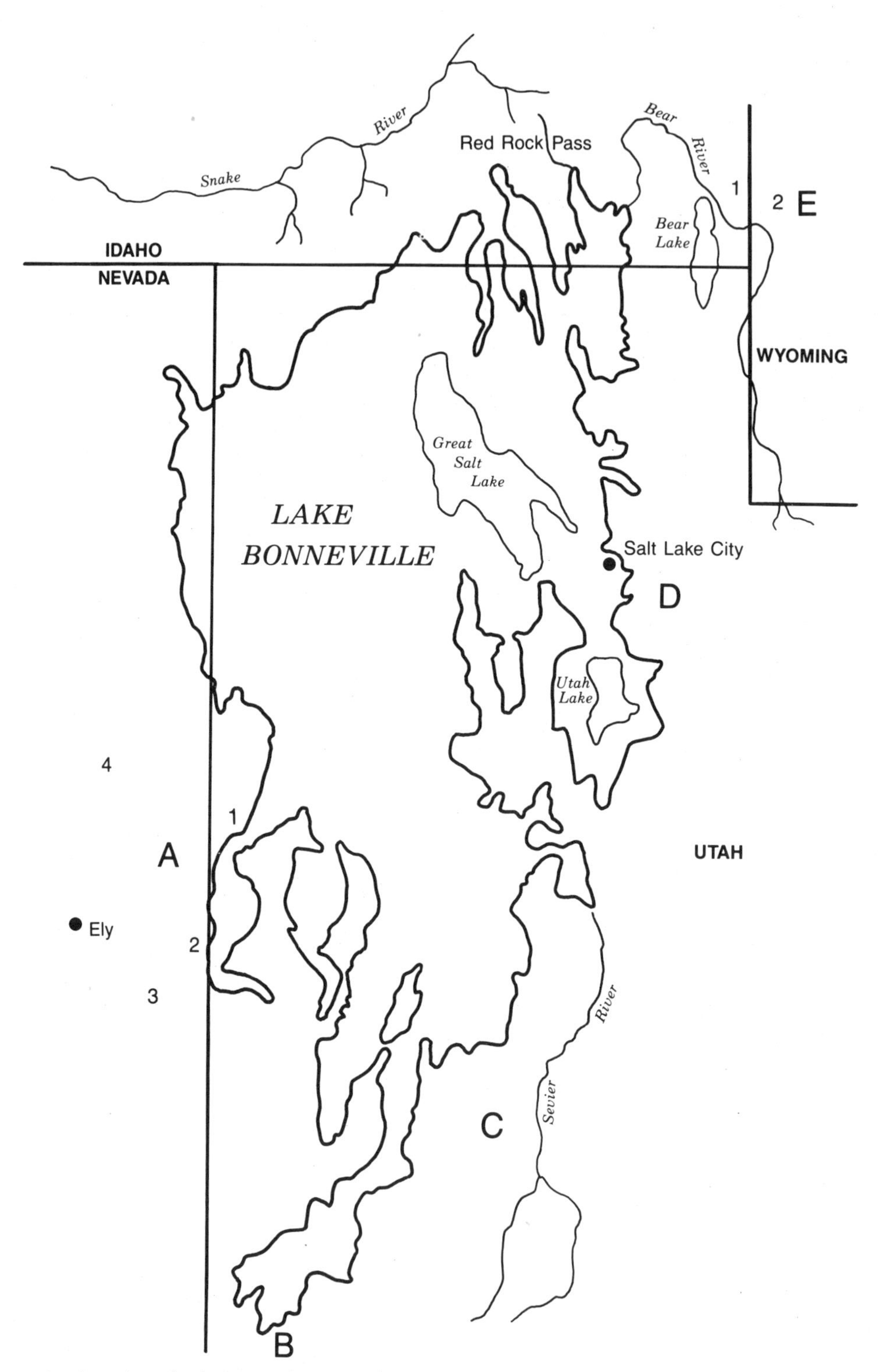

FIGURE 1.—Locations, by drainage, of pure Bonneville cutthroat trout populations (revised from Hickman 1978). **A. Utah-Nevada.** *Snake Valley drainage*: Trout and Birch creeks, Juab County, Utah (1); Hendry's and Hampton creeks, White Pine County, Nevada (2). *Spring Valley drainage*: Pine and Ridge creeks, White Pine County, Nevada

Department, USFS, and BLM. Four other populations have been identified as essentially pure. The seven pure populations occur in about 28 km of stream and in one 93-hectare lake. The Wyoming Game and Fish Department has an active management program that includes maintenance of brood stock at the Daniel Hatchery and the annual production of 60,000 fingerlings for stocking. Both the USFS and the BLM have managed grazing and watershed programs to enhance stream riparian habitat.

Utah

Twenty-one pure populations occur on three national forests and BLM public land in Utah. These occur in approximately 151 km of stream and one lake (Bear Lake). The Utah Division of Wildlife Resources has developed a draft Native Cutthroat Trout Management Plan for cooperative management of the subspecies with the USFS and BLM. National Forest plans and the BLM afford the subspecies "sensitive" status. Instream habitat improvements and livestock grazing systems have enhanced riparian habitats.

National Forest administrators are using the Bonneville cutthroat trout as a "management indicator species" to monitor the relative health of the aquatic–riparian ecosystem. Using the General Aquatic Wildlife System, they collect information on stream condition and calculate a habitat condition index (HCI). This HCI reflects habitat quality through such variables as pool quality, pool:riffle ratio, stream bottom materials, and streambank soil and vegetation stability. Managers in the Wasatch–Cache National Forest, for example, have established a current HCI value from inventory data, then expanded that value to a potential habitat capability HCI value. This value will be the goal of planned and future forest land management activities. This system is being used in the other Great Basin states to manage the Bonneville cutthroat trout habitat condition.

Nevada

Five genetically pure populations have been identified in Nevada adjacent to the Utah border. Two of these populations occur within the Bonneville Basin and three occur immediately outside the basin near Ely, Nevada, apparently a result of past transplanting efforts.

The Nevada Department of Wildlife prepared a Bonneville Cutthroat Trout Management Plan, in 1986, that defined the current status of the subspecies and prescribed a management and enhancement program. This plan identified the five streams, totaling 36 km, that support Bonneville cutthroat trout. These streams had been affected in 1983–1984 by unusual flood events. The department classifies the Bonneville cutthroat trout as a game fish and a sensitive species with priority for perpetuation and enhancement. A major portion of the streams supporting Bonneville cutthroat trout are in the Humboldt National Forest where they are considered sensitive species and accorded management indicator status in the recently completed Land and Resource Management Plan. One stream is on BLM public land and is being managed as sensitive species habitat by the Ely District of that agency.

Habitat conditions are considered fair in all five streams despite recent flood damage. Stream and riparian habitat improvements have been accomplished in several of the streams, and grazing is being controlled to limit riparian impacts. Generally, there have been improvements in population and habitat conditions in Nevada.

Future Species-Enhancement Plans

The past, current, and future management needs of the Bonneville cutthroat trout and its

FIGURE 1.—Continued. (3). *Steptoe Valley drainage*: Goshute Creek, White Pine County, Nevada (4).

B. Southern Utah. *Virgin River drainage*: Water Canyon, Reservoir Canyon, Leap, South Ash, Mill, and Harmon creeks, Washington County.

C. South-central Utah. *Sevier River drainage*: Birch, Pine, and North Fork North creeks, Beaver County; Sam Stowe Creek, Sevier County; Deep Creek, Garfield County.

D. Northern Utah. *Jordan River drainage*: North Fork Deaf Smith Canyon and Red Butte creeks, Salt Lake County. *Weber River drainage*: Moffitt Creek, Summit County. *Bear River drainage*: Bear Lake, Rich County; Carter, Meadow, McKenzie, and Sugarpine creeks, Summit County.

E. Idaho-Wyoming. *Bear River-Thomas Fork drainage*: Upper Giraffe, Pruess, and Dry creeks and Bear Lake, Bear Lake County, Idaho (1). Upper Giraffe*, Middle Giraffe, Raymond*, Upper Coal*, Huff, Salt, and Water Canyon* creeks, Lincoln County, Wyoming (2). *Bear River-Smith Fork drainage*: Coantag-Hobble*, Coal (Howland)-Sawmill*, and Porcupine creeks and Lake Alice*, Lincoln County, Wyoming (2).

Asterisks (*) denote Wyoming populations with a "Grade A" purity rating (Table 2). Other Wyoming populations have an essentially pure "Grade B" rating.

habitat were reviewed by four state wildlife agencies and two federal land management agencies during an interagency conference on the subspecies held in Salt Lake City, Utah, in March 1987. Convened by the U.S. Fish and Wildlife Service (USFWS), the conference provided a current status report on subspecies. Future activities, planned by state, are as follows.

Idaho

Future activities in Idaho will be directed toward riparian habitat enhancement in the Caribou National Forest and monitoring of existing populations by the Idaho Department of Fish and Game. No plans have been made to expand population range.

Nevada

Emphasis in Nevada will be placed on monitoring habitat and population numbers in the streams presently inhabited by Bonneville cutthroat trout, and these populations will be used as a source for future reintroductions. Plans include eradication of nonnative fish populations and reintroduction of the subspecies in six streams in its natural range between now and 1992. Efforts will be made to establish populations outside the subspecies' native range in four closed-basin streams. Evaluation of sportfishing potential of the subspecies will be conducted.

A major change in land management status took place in 1987 when the responsibility for management of about 77,000 acres was transferred from the Humboldt National Forest to the National Park Service (NPS) for the new Great Basin National Park near Ely, Nevada. Headwater areas of three streams in the Mt. Wheeler Peak area now come under the responsibility of NPS. The NPS has pledged to continue management efforts with the Nevada Department of Wildlife and USFS for the subspecies. The BLM and the USFS have assigned "sensitive species" status to the Bonneville cutthroat trout in their management plan and will continue to monitor habitat conditions and attempt to improve riparian habitat conditions. These agencies have implemented in-stream habitat improvements and changed grazing systems to reduce livestock impacts to riparian areas.

Utah

A major program that includes assessing the potential of this cutthroat subspecies for a variety of sportfishing management objectives as well as expanding the range of pure stocks has been initiated in Utah. A brood-fish program has been established for the Bear Lake cutthroat trout strain that involves a cooperative program between the Utah and Idaho agencies to collect adults from natural runs in two tributary streams of Bear Lake (Nielsen and Lentsch 1988). A majority of resulting hatching offspring will be returned to the lake, but a portion of the annual production will be used for other management programs. Projects are underway to establish Bonneville cutthroat trout in at least six streams and to establish populations in two reservoirs; brood fish from the reservoirs will be used for special management programs.

Special regulations (closures) have been implemented on two streams. All other streams are open to angling, but they are in remote locations and their fisheries are unpublicized.

The USFS and the BLM have prescribed several stream and riparian habitat improvement programs for the streams supporting Bonneville cutthroat trout. These agencies will monitor habitat conditions and grazing impacts on these streams.

Wyoming

The primary emphasis in Wyoming is to enhance Bonneville cutthroat trout populations through habitat improvement of streams known to support the Bonneville cutthroat trout. The BLM, USFS, and Wyoming Game and Fish Department have programs to improve riparian and aquatic habitats through grazing management, vegetative planting, beaver management, and the construction of instream structures. The Wyoming Game and Fish Department has an active stream habitat improvement program that has helped rehabilitate and restore many streams. Some supplemental stocking is planned by the department for streams in the Bonneville Basin, and efforts to improve genetic quality of the brood stocks are planned at the Daniel Hatchery. Angling is allowed in all streams; however, special regulations have been set in three streams since 1984. These regulations deal with number, size, and artificial lure restrictions.

Interagency Recommendations

As a result of the March 1987 Interagency Conference, responsible state and federal agencies concluded that land use practices are still the most important factors affecting overall condition of existing populations of Bonneville cutthroat trout. There is a limited number of problems that threaten specific populations, but the wide distribution of small, isolated populations limit the likelihood that any specific action will endanger a major area of the subspecies' current distribution. Introduction of nonnative trouts no longer appears to be a major threat because of policies that mandate preservation of current populations, identification of unknown populations, and expansion of the subspecies' range.

As a result of the Interagency Conference, the USFWS has concluded that significant advances have been made in negating threats to the continued existence of the Bonneville cutthroat trout (Anonymous 1987). There also appears to be an important commitment by all resource agencies to continue to work toward enhancing the existing population through several land and fisheries enhancement measures. The USFWS is recommending that the current threatened listing package be withdrawn indefinitely and the subspecies reclassified as a Category 2 candidate species (no need to list pending additional information). If a decline in or major threat to the subspecies or diminishing progress towards its enhancement occurs, its status will be reevaluated. Managers hail the interagency recognition that this subspecies may be potentially valuable to fisheries management but caution that it and other strains could be adversely affected unless judicious culture and distribution programs are carried out. Widespread distribution of this subspecies outside its native range is not encouraged.

Management opportunities for the Bonneville cutthroat trout, given appropriate conditions, are intriguing because this subspecies may provide a self-sustaining fishery in habitats unsuited for other salmonids. Through dedicated interagency management efforts, the present and future preservation of this species should be secure within the Bonneville basin.

References

Anonymous. 1987. Interagency Bonneville cutthroat conference. U.S. Fish and Wildlife Service, Summary Report, Salt Lake City, Utah.

Behnke, R. J. 1973. Status report: Utah or Bonneville cutthroat trout, *Salmo clarki utah*. Report to U.S. Fish and Wildlife Service, Albuquerque, New Mexico.

Behnke, R. J. 1976. Summary of information on the status of the Utah or Bonneville cutthroat trout, *Salmo clarki utah*. Report to U.S. Forest Service, Wasatch-Cache National Forest, Salt Lake City, Utah.

Behnke, R. J. 1979. The native trouts of the genus *Salmo* of western North America. Report to U.S. Fish and Wildlife Service, Denver, Colorado.

Behnke, R. J. 1980. Purity evaluation of Bear River cutthroat trout from Mill and Carter creeks, Wasatch National Forest, Summit County, Utah. Report to the U.S. Forest Service, Wasatch-Cache National Forest, Salt Lake City, Utah.

Behnke, R. J., and M. Zarn. 1976. Biology and management of threatened and endangered western trouts. U.S. Forest Service General Technical Report RM-28.

Binns, N. A. 1977. Present status of indigenous populations of cutthroat trout, *Salmo clarki*, in southwest Wyoming. Wyoming Game and Fish Department, Fisheries Technical Bulletin 2, Cheyenne.

Binns, N. A. 1981. Bonneville cutthroat trout, *Salmo clarki utah*, in Wyoming. Wyoming Game and Fish Department, Fisheries Technical Bulletin 5, Cheyenne.

Hickman, T. J. 1978. Systematic study of the native trout of the Bonneville basin. Master's thesis. Colorado State University, Fort Collins, Colorado.

Holden, P. B., W. White, G. Sommerville, D. Duff, R. Gervais, and S. Gloss. 1974. Threatened fishes of Utah. Proceedings of the Utah Academy of Science, Arts, and Letters 51(2):46–55.

Loudenslager, E. J., and G. A. E. Gall. 1980. Biochemical systematics of the Bonneville Basin and Colorado River cutthroat trout. Report to the Wyoming Department of Game and Fish, Cheyenne.

May, B. E., J. Leppick, and R. Wydoski. 1978. Distribution systematics and biology of the Bonneville cutthroat trout, *Salmo clarki utah*. Utah Division of Wildlife Resources, Publication 78-15, Salt Lake City.

Martin, M. A. 1982. The electrophoretic analysis of cutthroat trout subspecies in selected Utah waters. Master's thesis. Brigham Young University, Provo, Utah.

Miller, R. R. 1972. Threatened freshwater fishes of the United States. Transactions American Fisheries Society 101:239–252

Nielson, B. R., and L. Lentsch. 1988. Bonneville cutthroat trout in Bear Lake: status and management. American Fisheries Society Symposium 4:128–133.

Suckley, G. 1874. On the North American species of salmon and trout. Pages 91–160 *in* Report of the Commissioner for 1872 and 1873. U.S. Fish Commission and Fisheries, part 2, Washington, D.C.

Thompson, C. D. 1987. Enzyme electrophoresis of selected Utah cutthroat trout populations. Report to Utah Division of Wildlife Resources, Logan.

Thwaits, R. G. 1907. Early western travels, 1748–1846, volume 21. Arthur H. Clark Company, Cleveland, Ohio.

American Fisheries Society Symposium 4:128–133, 1988

Bonneville Cutthroat Trout in Bear Lake: Status and Management

BRYCE R. NIELSON

Utah Division of Wildlife Resources, Post Office Box 231, Garden City, Utah 84028, USA

LEO LENTSCH

Utah Division of Wildlife Resources, Post Office Box 349, Heber, Utah 84032, USA

Abstract.—The Bear Lake population of Bonneville cutthroat trout *Salmo clarki utah* evolved along with four endemic species of whitefish (*Prosopium*) and sculpin (*Cottus*) in Bear Lake, Utah–Idaho, during the past 28,000 years. Original populations were subjected to human activities in the early 1900s that resulted in a gradual population decline and, possibly, changes in genotype. The Bear Lake cutthroat trout was thought to have become extinct or to have hybridized, but recent research projects conducted by the Utah Division of Wildlife Resources have increased and enhanced the existing population and have provided descriptions of its meristic and biochemical characteristics. Results of these projects indicate that the existing Bear Lake cutthroat trout is a genetically pure, lacustrine form of the Bear River Bonneville cutthroat trout. Bear Lake cutthroat trout are spring spawners that mature late and attain large sizes; they are primarily piscivorous and are long lived. Intensive management of Bear Lake cutthroat trout is continuing. Recent introductions of hatchery-reared Bear Lake cutthroat trout into Strawberry and Lost Creek reservoirs in Utah and Blackfoot Reservoir in Idaho have provided preliminary data that indicate their applicability for management in other waters.

Bear Lake, bisected by the Utah–Idaho border, is a large, oval, natural lake within the Bear River drainage of the Bonneville Basin. Located at an elevation of 1,805 m, it is 32 km long and 13 km wide, and has a surface area of 282 km^2. Classified as an oligotrophic lake, it has had a continuous lacustrine existence for at least 28,000 years (Robertson 1978). Three important natural tributaries enter the lake: Swan, St. Charles, and Big Spring creeks; however, the largest inflow occurs from the Bear River which is diverted, via canals and water control structures, through Dingle Marsh into Bear Lake (Figure 1).

The ecosystem was isolated from any major drainages prior to the diversion of the Bear River system into the lake for water storage in the early 1900s. At least four endemic fish species occupy Bear Lake: Bonneville cisco *Prosopium gemmiferum*, Bear Lake whitefish *Prosopium abyssicola*, Bonneville whitefish *Prosopium spilonotus*, and Bear Lake sculpin *Cottus extensus*. These species provided forage for the indigenous Bear Lake population of Bonneville cutthroat trout *Salmo clarki utah*. Early accounts of this native cutthroat trout population are sketchy, but apparently populations were extensive enough to attract Indian tribes to the area and to provide a small commercial fishery at end of the 19th century (McConnell et al. 1957).

The first scientific observation of the Bear Lake cutthroat trout was made by G. Kemmerer and W. R. Boorman as part of a preliminary examination of western trout waters. They reported that "large numbers of bluenose trout *Salmo virginalis* and Williamson's whitefish are taken from the lake." Elsewhere they mentioned that the stomach of one bluenose trout "was well filled with smaller fish, which proved to be whitefish *Coregonus williamsoni*" (Kemmerer et al. 1923). Other early accounts include a newspaper article that discussed the existence of a rare bluenose trout that was found in exceptionally deep water, had a deep blue cast, and was unavailable to anglers.

During this period, early settlers of the Bear Lake Valley diverted Bear Lake tributaries for agricultural activities. Millions of Bear Lake cutthroat trout eggs were collected in the 1930s for stocking into other Utah and Idaho waters, and only small numbers were returned to Bear Lake. Rainbow trout *Salmo gairdneri* and lake trout *Salvelinus namaycush* were also introduced. As a result, the Bear Lake cutthroat trout population began a gradual decline. The collection of eggs from resident fish ceased in 1954, and stocking of cutthroat trout was limited to juvenile plants from other egg sources. Until recently, the indigenous Bear Lake cutthroat trout was thought to be extinct (Hickman 1978).

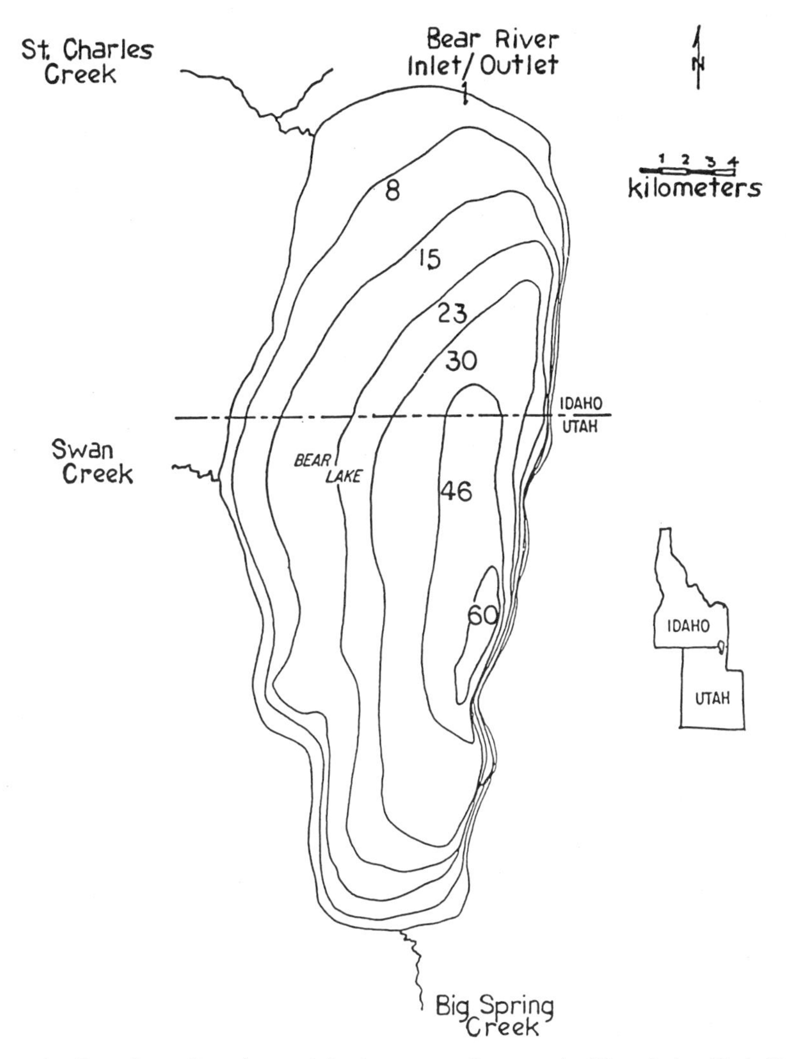

FIGURE 1.—Location, tributaries, and depth contours (in meters) of Bear Lake, Utah–Idaho.

Recognizing the need for a different management approach, the Utah Division of Wildlife Resources (UDWR), with the cooperation of the Idaho Department of Fish and Game (IDFG), initiated a Federal Aid in Fish Restoration project in 1973 in an attempt to increase and enhance the remaining population and gene pool of the Bear Lake cutthroat trout through fish culture and intensive management. A new coldwater hatchery was constructed to rear Bear Lake cutthroat trout at Mantua, Utah. Approaches taken included the collection and culture of eggs from wild Bear Lake cutthroat trout, marking and stocking the resultant fish at appropriate sizes and times, evaluating the year-class survival and life history data through gill-net surveys, and evaluating the sport fishery and the contribution of stocked fish to the creel. This effort has expanded and continues to the present time.

Taxonomy

Phenotypically, the Bear Lake cutthroat trout are silver with minimal orange or yellow coloration on the body. Maxillary slashes are not typical, and they are faint if present. The pectoral and pelvic fins exhibit distinctive yellow coloration, and single, random, intense-yellow splotches may be found on some individuals. Adult fish develop an azure-blue coloration, which extends from the snout along the back and becomes more pronounced shortly after death; when spawning time approaches, the azure-blue coloration disappears.

Males develop a rose-orange color, especially on the ventral regions and on the opercles, and females are a drab olive-gray or silver color. Spotting patterns of both sexes consist of irregular black spots dispersed above the lateral line and behind the dorsal fin, with increased concentrations posteriorly. Spots are rarely found on the head or on the back anterior to the dorsal fin.

McConnell et al. (1957) stated that the pure *Salmo clarki utah* originally found in Bear Lake had been replaced by hybrids formed between itself and rainbow trout or Yellowstone cutthroat trout *Salmo clarki bouvieri*, or both. Behnke (1979), however, stated that the cutthroat trout of Bear Lake exhibited only minute hybridization and represented a genotype of the original lacustrine stock of Bonneville cutthroat trout from ancient Lake Bonneville.

Efforts were made to clarify the degree of hybridization of the existing Bear Lake cutthroat trout with rainbow trout and other cutthroat trout subspecies. Similarities to Bonneville cutthroat trout strains were also noted. Meristic and electrophoretic data were gathered from fish captured in gill nets and from wild adults whose eggs were collected for brood-stock replacement in 1981 through 1983.

A meristic description was made from a sample of 20 Bear Lake cutthroat trout ranging in age from age 2 to 6 that were obtained from gill nets in 1982. Counts were made of scales in the lateral-line series, scales above the lateral line, basibranchial teeth, gill rakers, and pyloric caeca. These meristic characters were selected because they are most useful for separating species and subspecies of trouts (Behnke and Zarn 1976). Analytical methods were similar to those suggested by Hubbs and Lagler (1967) and Hickman (1978). The number of pyloric caeca is the character least influenced by environmental changes, and is the most useful in separation of species and subspecies (Wernsman 1973). Data on counts of pyloric caeca and gill rakers from 450 Bear Lake cutthroat trout gillnetted in 1981 were included for comparison.

Occurrence of basibranchial teeth is one of the main characters that distinguish cutthroat trout from rainbow trout. Even a slight hybrid influence from rainbow trout is detected most readily in loss of basibranchial teeth (Behnke 1979). Hickman (1978) distinguished pure populations of *Salmo clarki* by the presence of basibranchial teeth in at least 90% of the population. In the 20 Bear Lake cutthroat trout examined in 1982, basibranchial

TABLE 1.—Comparison of meristic data between pure Bonneville cutthroat trout, summarized by Hickman (1978), and the Bear Lake population of cutthroat trout. N.A. = not available.

Measure	Statistic	Pure Bonneville cutthroat trout	Bear Lake cutthroat trout
Scales in lateral line	Range	133–183	150–183
	Mean	160	165
Scales above lateral line	Range	33–46	32–41
	Mean	38	37
Basibranchial teeth	Range	1–50	6–22
	Mean	N.A.	14
	Frequency of occurrence	Present in at least 90% of population	100%
Gill rakers	Range	16–24	15–20
	Mean	19	18
Pyloric caeca	Range	25–54	41–65
	Mean	35	52

teeth were present in all specimens, with an average of 14 per fish. Rainbow trout also have lower scale counts (120–140; Behnke 1979) in the lateral-line series than Bear Lake cutthroat trout (150–183).

Meristic characteristics of the Bear Lake cutthroat trout and the pure Bonneville cutthroat trout were similar with the exception of the pyloric caeca counts (Table 1). Behnke (1979) stated that there are three slightly different groups of *S. c. utah* and that the Bear River drainage group typically averaged more than 40 pyloric caeca, whereas the other two groups averaged 35. Bear Lake cutthroat trout had a mean pyloric caeca count of 52. Field counts of gill rakers and pyloric caeca made in 1981 yielded means of 18 gillrakers and 51 pyloric caeca.

Biochemical data were collected from 23 adult spawners and 29 gill-netted Bear Lake cutthroat trout 1983. All Bear Lake cutthroat trout were homozygous at the phenylalanyl-proline *Phap(100)* and sorbitol dehydrogenase *Sdh-2(100)* alleles; these alleles are both used to differentiate the Bear River Bonneville cutthroat trout strain from other strains. Based on the 90% or higher frequency of these cutthroat trout alleles, the cutthroat trout of Bear Lake originated from the Bear River Bonneville strain (M. Martin, UDWR, unpublished data).

All sampled cutthroat trout lacked indications of hybridization with rainbow trout. Each fish was homozygous, and all loci were fixed for the proteins, leucyl-glycyl-glycine (LGG), glycyl-lucine (GLP), and "malic" enzyme (ME), that are used to designate rainbow trout hybridization with cutthroat trout. Isocitrate dehydrogenase (IDH) expressed cutthroat trout alleles with no genetic variation in the Bear Lake population.

Life History

Spawning activities begin in late April and are completed by late June; UDWR and IDFG have operated spawning traps on Swan and St. Charles creeks since 1973 (Nielson 1981, 1986). Marking and tagging information indicated that Bear Lake cutthroat trout begin maturing at age 5 but may not initiate spawning until they are more than 10 years old. Females typically outnumber males, and fish commonly enter the tributaries on sunny afternoons. Mature fish average 560 mm total length (TL) and weigh approximately 2 kg. There was little difference in the sizes of males and females. The average age of 276 marked fish in the 1987 spawning run at Bear Lake was 6.8 years (range, age 4 to 11). Over 92% of the fish in the spawning run were age 6 or older. Maximum ages have not been determined, but fish weighing more than 8 kg have been trapped. Repeat spawners account for less than 4% of the total run.

No research has been done on the early life history of naturally spawned Bear Lake cutthroat trout. General observations, and an understanding of the ecosystem in which they evolved, indicated that juveniles would spend 1–2 years in the tributary before they entered the lake during spring runoff. This scenerio is based on the assumption that, because of the low productivity of Bear Lake, young fish compete with the endemic planktivores for the limited numbers of small-sized zooplankton. No data relating to competion or downstream migration have been collected because the majority of the spawning population is captured in spawning traps.

Bear Lake cutthroat trout inhabit the littoral and pelagic zones but are limited during summer stratification to depths at or below the thermocline. Stocked yearling (125 mm TL) cutthroat trout feed primarily on terrestrial insect drift in the spring and summer. Zooplankton are rarely found in stomach samples; however, fish are occasionally present. As the Bear Lake cutthroat trout grow and mature, they become primarily piscivorous, feeding on Bear Lake sculpins and Bonneville ciscoes (Nielson 1982); however, they are opportunistic predators and prey on all components of the available food base. Food habit data indicated that increases in total length were proportional to the incidence of fish in the diet. Between 1973 and 1981, 20% of the Bear Lake cutthroat trout averaging 250 mm TL were piscivorous, whereas 95% of the cutthroat trout over 550 mm TL consumed fish. Terrestrial insects

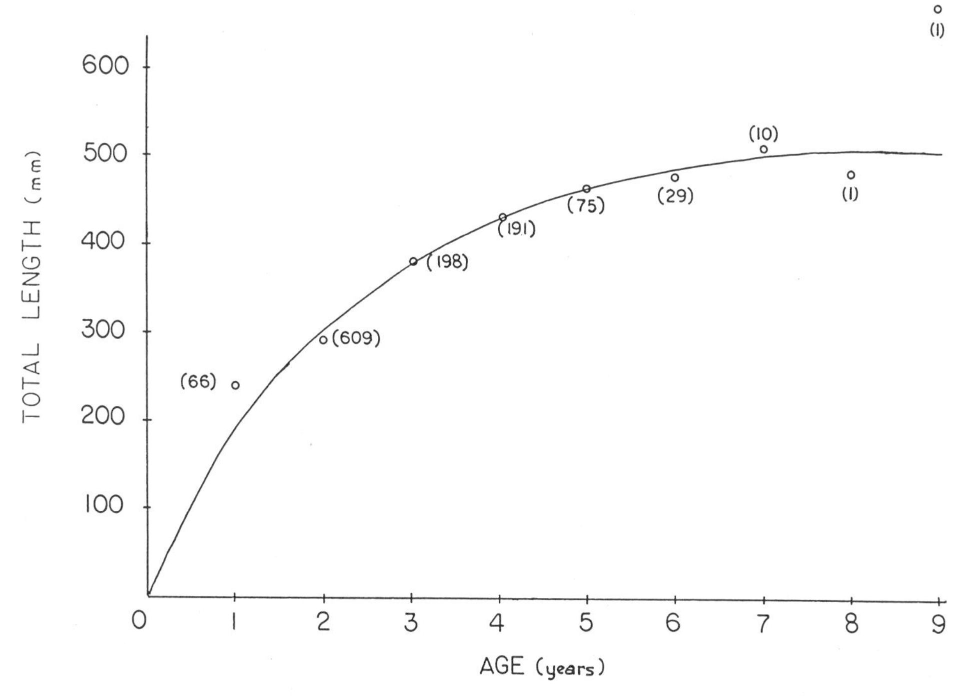

FIGURE 2.—Mean total lengths (mm) of gillnetted known-age cutthroat trout in Bear Lake, 1975–1985. Sample sizes are in parentheses. (From Nielson 1986.)

accounted for the remainder of the food base, and benthic invertebrates and zooplankton rarely occurred in the diet (Nielson 1986). Data on winter food habits are lacking, but examination of stomach contents of creeled fish indicated a reliance on whitefishes and their eggs.

Age and growth of cutthroat trout in Bear Lake have been documented over the past 10 years with marked, known-age year classes (Figure 2). Growth rates are slow, averaging approximately 50 mm TL annually, but because of their longevity and late maturity, individuals frequently exceeded 500 mm TL (Nielson 1986).

Status and Management

Bear Lake cutthroat trout are managed intensively on Bear Lake. Wild Bear Lake cutthroat trout are trapped and spawned yearly on Swan and St. Charles creeks. Through morphological selection of specific traits such as spotting patterns, body configuration, size, and the ability to survive to maturity in Bear Lake, efforts are made to maintain the genotype and to improve the spawning stock. Unfortunately, preexisting water rights on the spawning tributaries impede the restoration of natural spawning runs. Data are collected on the spawning dynamics and the contribution of various stocked cohorts. Wild eggs are also collected to replace and improve the existing brood stock that is maintained at the J. Perry Egan Hatchery near Bicknell, Utah. The resulting eggs from wild and captive fish are reared at Mantua Hatchery for reintroduction into Bear Lake.

All fish are marked with fluorescent grit or a fin-clip, or both, prior to stocking. Various stocking density, size, and timing scenerios have been employed to determine the optimum management approach for maximum survival on a cost-efficient basis. These have included spring stockings of fish (125 mm TL) at an average rate of 500,000 fish/year (1976–1980); spring stockings of fish (200 mm TL) at an average of 160,000 fish/year (1981–1983); fall stockings of fish (175 mm TL) at an average of 400,000 fish/year (1981–1983); and spring stockings of fish (150 mm TL) at an average of 1,000,000 fish/year (1984–1986). Standardized gillnetting has yielded data on poststocking survival, food habits, age and growth, distribution, meristic characters, and condition factors. Creel surveys have provided information to help describe the fishery in terms of catch rates, mark returns, angling pressure, and harvest. Total harvest has increased from 500 fish in 1973 to over 14,000 in 1985 (Nielson 1986). At present the Bear Lake cutthroat trout population has been restored and is continuing to expand slowly. Fishing pressure and total harvest have increased; however, since 1979, annual catch rates for cutthroat trout have been stable at 0.08 fish/h. Vulnerability to angling is highest during the winter through the ice. Gill-net catch rates have remained constant since 1981 at 0.10 fish/net-hour. Marked fish have accounted for 88% of the catch of cutthroat trout in the gill nets during this period.

Uniformity in the gill-net and creel survey data during the study years existed despite different stocking rates and fish sizes. Success of the spring stockings varied and is still being assessed, but fish stocked in the fall apparently failed because they were infrequently observed in gill nets or creels (Nielson 1986). Future research at Bear Lake will include studies on attracting stocked spawners with morpholine, poststocking survival, predator–prey relationships, distribution and food habits in terms of depth on a year-round basis, and benthic invertebrate and zooplankton dynamics.

Preliminary results from the Bear Lake Cutthroat Project have elicited substantial interest in utilizing Bear Lake cutthroat trout in other waters. Specific attributes of this fish include a piscivorous nature, vulnerability to angling, attainment of large size at maturity, a potential to reach trophy size, ability to coexist with efficient planktivores, wide tolerance to various habitats, high reproductive potential, and an ability to maintain reproductive isolation. Fish reared from excess brood-stock eggs have been used for introduction into other waters.

In 1983, IDFG introduced Bear Lake cutthroat trout into Blackfoot Reservoir, an irrigation impoundment on the Blackfoot River that covers 8,100 hectares and has a maximum depth of 11.6 m. An average of 150,000 fish (125 mm TL) have been stocked each spring since 1983. Fish were treated with morpholine prior to stocking. Bear Lake cutthroat trout began appearing in creels in 1985 and were a major portion of the catch by early 1986. Fish have been found in stomach samples from these cutthroat trout; however, the percentage of fish exhibiting piscivorous behavior has not been estimated. The first return of spawners was in 1986, at age 4, when 145,000 eggs were taken, but large-scale production did not develop until 1987 when approximately 700,000 eggs were taken (J. Heimer, IDFG, personal communication).

Bear Lake cutthroat trout were also introduced into two coldwater irrigation impoundents in Utah, Strawberry and Lost Creek reservoirs.

Strawberry Reservoir is a high-elevation (2,495 m) mesotrophic reservoir presently being enlarged from its original 1,336 hectares to 4,850 hectares. Bear Lake cutthroat trout were stocked as part of a program of strain evaluation to determine which cutthroat trout subspecies would exhibit the most desirable management attributes for the reservoir. Equal numbers (approximately 180,000; 125 mm TL) of marked Bear Lake cutthroat trout and Strawberry reservoir hybrids (*Salmo clarki bouvieri* × *Salmo gairdneri*) were stocked in the springs of 1985 and 1986. Performance was monitored through monthly gillnetting (May–November) and an intensive, season-long angler survey (May–November).

Lost Creek Reservoir supports rainbow trout and cutthroat trout fisheries and has a nongame fish population that is presently expanding after chemical renovation in 1977. Bear Lake cutthroat trout were introduced into this reservoir to determine their growth rates and food habits in a productive water body containing an expanding population of Utah chub *Gila atraria*. Approximately 10,000 marked Bear Lake cutthroat trout (125 mm TL) were stocked in May of 1986 and 1987. Their performance has been evaluated through gill-net surveys in June, August, and October.

The growth pattern of Bear Lake cutthroat trout in introduced waters was excellent, and a large proportion of annual growth occurred during the winter. In Strawberry and Lost Creek reservoirs, Bear Lake cutthroat trout annually grew 125 and 68 mm TL, respectively. The Strawberry cutthroat trout hybrids did not exhibit winter growth. Zooplankton was the most common food item in these productive reservoirs. Only 5% of the marked cutthroat trout were piscivorous in either water, but this result was influenced by the average size (<300 mm TL) of the fish sampled. The Bear Lake cutthroat trout converted to a fish diet at a smaller size (351 mm TL) than the Strawberry cutthroat trout (386 mm TL). Creel rates in Strawberry Reservoir for equal numbers of stocked cutthroat trout from the two groups in 1985 and 1986 were 0.055 and 0.004 fish/h, respectively, for Bear Lake cutthroat trout and 0.004 and 0.001 fish/h, respectively, for Strawberry cutthroat trout.

The UDWR remains committed to the maintenance, propogation, and distribution of the Bear Lake cutthroat trout. Availability of excess Bear Lake cutthroat trout for stocking elsewhere is extremely limited; however, efforts are being made to increase numbers of captive broodstock and to maximize production from wild populations. Management programs for use of Bear Lake cutthroat trout are focused on reservoirs with substantial angling pressure, good forage fish populations, and natural reproduction potential. The value of Bear Lake cutthroat trout in stream fisheries has not yet been addressed. These fish are not expected to exhibit superior performance in all waters, but when specific traits are desirable, management possibilities are interesting.

References

Behnke, R. J. 1979. The native trouts of the genus *Salmo* of western North America. Report to the U.S. Fish and Wildlife Service, Denver, Colorado.

Behnke, R. J., and M. Zarn. 1976. Biology and management of threatened and endangered western trout. U.S. Forest Service General Technical Report RM-28.

Hickman, T. J. 1978. Systematic study of the native trout of the Bonneville basin. Master's thesis. Colorado State University, Fort Collins.

Hubbs, C. L., and K. F. Lagler. 1967. Fishes of the Great Lakes region. University of Michigan Press, Ann Arbor.

Kemmerer, G., J. F. Bovard, and W. R. Boorman. 1923. Northwestern lakes of the United States; biological and chemical studies with reference to possibilities in production of fish. U.S. Bureau of Fisheries Bulletin 39:51–140.

McConnell, W. J., W. J. Clark, and W. F. Sigler. 1957. Bear Lake, its fish and fishing. Utah Department of Fish and Game, Idaho Department of Fish and Game, and Wildlife Management Department, Utah State Agricultural College, Logan.

Nielson, B. R. 1981. Bear Lake cutthroat enhancement program, 1975–79. Utah Division of Wildlife Resources, Federal Aid in Fish Restoration, Project F-26-R-5, 5-year Completion and 1979 Annual Performance reports, Salt Lake City.

Nielson, B. R. 1982. Bear Lake cutthroat trout enhancement program, 1981. Utah Division of Wildlife Resources, Federal Aid in Fish Restoration, Project F-26-R-7, 1981 Annual Performance Report, Salt Lake City.

Nielson, B. R. 1984. Bear Lake cutthroat trout enhancement program, 1983. Utah Division of Wildlife Resources, Federal Aid in Fish Restoration, Project F-26-R-9, 1983 Annual Performance Report, Salt Lake City.

Nielson, B. R. 1986. Bear Lake cutthroat trout enhancement program, 1980–84. Utah Division of Wildlife Resources, Federal Aid in Fish Restoration, Project F-26-R-10, 5-year Completion and 1984 Annual Performance reports, Salt Lake City.

Robertson, G. C. 1978. Surficial deposits and geologic history, northern Bear Lake valley, Idaho. Master's thesis. Utah State University, Logan.

Wernsman, G. 1973. The native trouts of Colorado. Master's thesis. Colorado State University, Fort Collins.

American Fisheries Society Symposium 4:134–140, 1988

Review of Competition between Cutthroat Trout and Other Salmonids

J. S. Griffith

Department of Biological Sciences, Idaho State University, Pocatello, Idaho 93209, USA

Abstract.—Most interior subspecies of cutthroat trout *Salmo clarki*, such as the Yellowstone cutthroat trout *S. c. bouvieri* in Yellowstone National Park, evolved in the presence of few other fish species. Consequently, they have developed as "generalists" whose behavior and morphology may have been more thoroughly shaped by the environmental conditions under which they evolved than by interspecific interactions. The westslope cutthroat trout *S. c. lewisi*, which evolved with anadromous salmonids, may have developed a more specialized niche. Coastal cutthroat trout *S. c. clarki* evolved in a community more complex than that of other subspecies. When nonnative salmonids were introduced into their habitats beginning about 100 years ago, some cutthroat trout populations declined precipitously. Although competition was a factor in those declines, the mechanisms are still not well understood. Cutthroat trout populations that are affected by changes in habitat quality or are removed by anglers may be replaced by nonnative trouts.

The decline of cutthroat trout *Salmo clarki* populations in the past century has been described in many papers in this volume. This decline is at least partially attributed to interactions of cutthroat trout with other fish species, although detailed accounts of such interactions are generally limited. There are several types of interactions that may affect cutthroat trout. One is predation on cutthroat trout by introduced fish species; this may be substantial, especially in lakes (Marnell 1988, this volume). Another factor is genetic introgression, which has also had a great effect on cutthroat trout (Behnke 1979). Neither of these factors will be discussed here.

Another category of interactions is interspecific competition. According to Birch (1957), "Competition occurs when a number of animals (of the same or different species) use common resources the supply of which is short; or if the resources are not in short supply, competition occurs when the animals seeking that resource nevertheless harm one another in the process." The objectives of this paper are to briefly review the extent of interactions between cutthroat trout subspecies and the other fish species with which they evolved and to examine evidence of competition between cutthroat trout and introduced salmonids.

Interactions with Native Species

Too little emphasis has generally been placed on identifying the fish species with which cutthroat trout evolved or on understanding the interactions between them. Cutthroat trout subspecies and individual populations show a gradation in the complexity of these interactions. At one extreme is the population of Yellowstone cutthroat trout *Salmo clarki bouvieri* in Yellowstone Lake (Yellowstone National Park), where longnose dace *Rhinichthys cataractae* was the only other native fish species present (Gresswell and Varley 1988, this volume). Of the 12 native fish species in Yellowstone National Park, only mountain whitefish *Prosopium williamsoni* and Arctic grayling *Thymallus arcticus* evolved in niches vaguely similar to that of the cutthroat trout (Varley and Schullery 1983).

Other populations of interior cutthroat trout, such as the Yellowstone cutthroat trout outside of the Yellowstone Lake area and the Great Basin subspecies, evolved among a relatively depauperate fish fauna, consisting of species of *Cottus, Catostomus*, and *Rhinichthys*, that is typical of much of North America (Moyle and Vondracek 1985). In the Lahontan Basin (Nevada–California), for example, the original fish fauna consisted of only eight species of catostomids, cyprinids, and cottids in addition to mountain whitefish and the Lahontan cutthroat trout *S. c. henshawi* (La Rivers 1962).

Westslope cutthroat trout *S. c. lewisi* appear to have been influenced by more substantial interactions with salmonids and with nonsalmonids such as the longnose sucker *Catostomus catostomus* and, in the Columbia River drainage, northern squawfish *Ptychocheilus oregonensis* and redside shiner *Richardsonius balteatus*. Westslope cutthroat trout evolved along with mountain whitefish and bull trout *Salvelinus confluentus*, and there is evidence of substantial segregation for food and space, with the latter two species being more benthic oriented (Pratt 1984). Westslope cutthroat are also naturally sympatric with chinook salmon *Oncorhynchus tshawytscha* and with rainbow trout and steelhead *Salmo gairdneri* in the part of their range that includes the Salmon and Clearwater rivers in Idaho (Behnke 1988; Liknes and Graham 1988, both this volume).

Hanson (1977) examined distribution and behavior of steelhead and cutthroat trout in central Idaho and found no sympatric fluvial populations. In drainages where both were found, cutthroat trout occupied upper portions of the streams, and steelhead were found in the lower portions. In experimental stream channels, age-0 steelhead were capable of displacing previously established age-0 cutthroat trout, but the latter could not displace steelhead. Hanson suggested that interaction between the two is ongoing and reflects interactive segregation in those streams, despite the fact that the species apparently have been coexisting since the last glacial epoch.

Moffitt and Bjornn (1984) examined changes in fish populations after the construction of Dworshak Dam blocked the anadromous steelhead from the North Fork Clearwater River in Idaho. No anadromous fish spawned there after 1969. Although densities of rainbow trout and steelhead decreased from about 18 fish/100 m^2 of stream before 1969 to 5 or fewer resident rainbow trout per 100 m^2 in 1974–1983, cutthroat trout densities did not increase correspondingly. About 2 cutthroat trout/100 m^2 were found in 1968, and this density existed in 1983, after having temporarily doubled during the mid-1970s (Moffit and Bjornn 1984). Failure of the cutthroat trout population to expand and fill habitat once occupied by juvenile steelhead may indicate that niches of the two species have become genetically differentiated through selective segregation.

Coastal cutthroat trout *Salmo clarki clarki* have evolved under the influence of more substantial interspecific interactions than the interior subspecies. Relationships of coastal cutthroat trout with rainbow trout and steelhead and with coho salmon *Oncorhynchus kisutch* have been extensively investigated. Where they inhabited the same watersheds in southwestern British Columbia, rainbow trout and steelhead occupied the larger river reaches, and cutthroat trout were more abundant in headwater tributaries (Hartman and Gill 1968). In agonistic encounters between individuals, rainbow trout and steelhead were behaviorally dominant over cutthroat trout. A similar segregation of juveniles occurred in Alaska streams (Jones 1978; Johnson et al. 1986). In Oregon, Nicholas (1978) found that coastal cutthroat trout in streams of the western Cascade range lived higher in the drainage, grew more slowly, and matured earlier than did sympatric resident rainbow trout. Cutthroat trout spawned earlier in the spring and in smaller or different tributaries than rainbow trout. Nicholas felt that such resource partitioning worked to maintain the integrity of these cutthroat trout, whereas in inland Oregon the cutthroat trout populations developed in allopatry and have now been replaced over much of their range by introduced rainbow trout.

Resource partitioning by coastal cutthroat trout and juvenile coho salmon occurs in small coastal streams on Vancouver Island, British Columbia (Glova 1984), and also occurred under laboratory conditions (Glova 1986). Coho salmon were larger in size than cutthroat trout, and their biomass was greater in lower stream reaches; cutthroat trout biomass was greater in the upper reaches. Coho salmon were more abundant in pools and glides, and cutthroat trout were numerically dominant in riffles. Some difference in their diets was evident, with adult insects, mainly Diptera and Hemiptera, being more common in coho salmon stomachs, and chironomid larvae and pupae were more common in cutthroat trout stomachs.

In lakes, coastal cutthroat trout display moderate resource partitioning with other native salmonids and show allopatric–sympatric shifts in habitat utilization and diet. Nilsson and Northcote (1981) examined food, size, and growth of cutthroat and rainbow trout in 27 coastal British Columbia lakes. In allopatry, rainbow trout fed extensively on benthic, midwater, and surface prey, and cutthroat trout utilized mostly midwater prey, especially fish when they were available. In sympatry, rainbow trout used mainly limnetic surface and midwater prey, and cutthroat trout used more littoral prey and were much more piscivorous. In laboratory tanks, rainbow trout were consistently more aggressive and they quickly killed cutthroat trout when they were paired together. Rainbow trout displayed different patterns of threat behavior and foraging behavior than did cutthroat trout. Sympatric lake populations of coastal cutthroat trout and Dolly Varden *Salvelinus malma* also showed a segregation of habitat and diet (Andrusak and Northcote 1971).

Competition from Introduced Salmonids

Competition among fluvial salmonids usually translates into attempts by individuals of the same or different species to secure territories for adequate space and, therefore, food or cover or both (Chapman 1966; Fausch and White 1981). Continued interactions between two species should produce a niche shift in one or both species, or the extinction of one species, or a fluctuating coexistence as environmental conditions alternately favor one species over the other (Moyle and Vondracek 1985). Hearn (1987) pointed out that populations that are limited by factors such as predation or by stochastic processes such as droughts, floods, or severe winters do not provide an opportunity for interspecific competition to exist. It is possible that the "ecological crunch" hypothesis of Wiens and Rotenberry (1981) may be appropriate for many salmonid populations. During much of the time, population numbers are regulated by climatic factors, so resources such as food are abundant and interspecific competition is minimal. During episodes of resource limitation (ecological crunches), interspecific interaction is intense, and the niches of one or both species are modified as a result. For fluvial trout, interspecific competition should be most intense in late summer in those years when population levels are at or above "normal." At such times when space and food were minimal, niche shifts have been observed, for example, between juvenile brook trout *Salvelinus fontinalis* and Atlantic salmon *Salmo salar* (Gibson 1973). In lakes, trout populations are less likely to be regulated by abiotic events, and ecological crunches might be expected to occur more frequently than in streams. On the other hand, except during spawning, lacustrine salmonids would not be expected to be territorial, therefore reducing the opportunity for agonistic encounters between individuals of different species.

Limited information is available concerning annual variations in cutthroat trout populations and the mechanisms that regulate them. Some data were provided by Platts and Nelson (in press), who monitored population fluctuations (the maximum percentage difference for a series of annual population estimates) of cutthroat trout and associated salmonids in three Great Basin streams during 1975–1985, a period that included some severe climatic conditions. Two of the study streams contained only pure strains of Lahontan cutthroat trout, and their populations fluctuated 448% in one stream and 772% in the other during the period. The third stream contained Yellowstone cutthroat trout, brown trout *Salmo trutta*, and rainbow trout. One study section showed a 288% fluctuation in cutthroat trout numbers and another section, which had low cutthroat trout densities, showed 800%. Platts and Nelson suggested that these data, and those from six other study streams where cutthroat trout were absent, indicated that stochastic factors were more important than competition or predation; fluctuations of allopatric salmonid populations were similar to, or greater than, those of populations in multispecies situations.

There are several aspects of the life history, behavior, and morphology of cutthroat trout that indicate a vulnerability to competition with other salmonids. Adult cutthroat trout may mature one or more years later than some of their competitors, especially brook trout. Cutthroat trout fry emerge from the gravel later in the year than many of their competitors and, thus, age-0 cutthroat trout acquire a statistically significant length disadvantage (when compared with age-0 individuals of other species) that may continue throughout their lifetime (Griffith 1972). Such a size discrepancy may enhance resource partitioning and, therefore, minimize interspecific competition, but in times of ecological crunch cutthroat trout may be at a disadvantage if they cannot hold territories against larger competitors. Also, cutthroat trout are innately less aggressive than equal-sized competitors such as steelhead or rainbow trout (Hanson 1977; Nilsson and Northcote 1981) and, therefore, they cannot compensate behaviorally for their smaller size.

A lack of morphological specialization by cutthroat trout may help explain their displacement from pools by coho salmon and their domination by steelhead in riffles. Coho salmon and steelhead displayed morphological adaptations for maneuvering and holding position in pools and riffles, respectively, but cutthroat trout were less well adapted to pools than coho salmon or to riffles than steelhead (Bisson et al., in press). Furthermore, body size is related to behavioral dominance, and juvenile cutthroat trout in southwestern Washington streams weighed an average of 10% less than juvenile steelhead of equal length (Bisson and Sedell 1984).

Competition between Brook Trout and Cutthroat Trout

It appears that cutthroat trout populations are less likely to coexist with brook trout than with other nonnative salmonids. In low-gradient tributaries of the Shields River, Montana, Yellowstone cutthroat trout populations had declined, by 1985, to one-half to one-third of their 1974 abundance and, in the same period, brook trout increased to outnumber the cutthroat trout. (C. Clancy, Montana Department of Fish, Wildlife and Parks, personal communication.) In Yellowstone National Park, the introduction of brook trout has nearly always resulted in the disappearance of the cutthroat trout (Varley and Gresswell 1988, this volume).

Possible competition between westslope cutthroat trout and brook trout, which were originally stocked in the 1940s, was examined in small Idaho streams by Griffith (1972, 1974). Although cutthroat trout were 20 mm shorter than brook trout of the same age-group and some resource partitioning was evident, no mechanisms were found that could completely account for the increase in brook trout and the concommitant decline in cutthroat trout.

In 1979, 10 years after the initial study, the four study streams were resampled using the same procedures. On Crystal Creek, a stream that previously held sympatric populations, the riparian forest had been logged, and few fish remained. Habitat in the other stream with sympatric populations, Hoodoo Creek, had not been altered. The density of cutthroat trout in Hoodoo Creek had not changed from the average of 4 fish/100 m^2 observed in 1969, but brook trout density had decreased from 27 to 21 fish/100 m^2. Densities in streams with allopatric populations were virtually unchanged from the 18–20 fish/100 m^2 of each species recorded in 1969.

Microhabitat analysis indicated that, for age-3 and older fish, the interspecific difference in focal-point water velocity that had existed in 1969 had increased in 1979. Mean focal-point velocity decreased from 7.6 cm/s in 1969 to 4.7 cm/s in 1979 for age-3 and older brook trout, and increased from 10.3 cm/s in 1969 to 12.2 cm/s in 1979 for age-3 and older cutthroat trout. These changes were significant for both species, and the velocity values were significantly different (χ^2; $P < 0.05$) between species. The age-3 and older cutthroat trout found in Hoodoo Creek were in groups of two to four territorial individuals in pools. Stream velocities faced by younger fish living in sympatry had not changed since 1969, nor had water depths occupied by any age-group.

In Hidden Valley Creek, Colorado, Cummings (1987) evaluated interspecific competition for preferred positions, during declining flows of late summer, between greenback cutthroat trout *S. clarki stomias* and brook trout. Study sites were closed to angling. Competition between adult (>150 mm long) cutthroat trout and brook trout was not apparent. However, after brook trout were removed, juvenile cutthroat trout shifted to occupy different positions. Focal-point velocities and distance to nearest cover decreased significantly; analysis of covariance was used to account for declining flows. During years of lower flow, the stream margins rapidly dewatered and the microhabitat for age-0 cutthroat trout declined. Cummings thought that dewatering forced cutthroat trout to occupy positions in the main stream channel where they must both expend more energy and compete with age-0 brook trout sooner than during high-flow years.

Competition between Cutthroat Trout and Other Introduced Species

The decline of cutthroat trout in streams has been attributed to competition from introduced rainbow trout (Hickman and Duff 1978; Moyle and Vondracek 1985; Gerstung 1988, this volume) and brown trout (Moyle and Vondracek 1985; Gerstung 1988), but details are lacking. In Martis Creek, California, endemic Lahontan cutthroat trout have been replaced by brown and rainbow trouts despite availability of cutthroat trout recruits from upstream. Moyle and Vondracek (1985) thought that the two introduced species together play the same ecological role in Martis Creek previously played by the cutthroat trout. In the Henrys Fork of the Snake River, Idaho, local residents related that cutthroat trout virtually disappeared, within 8 years, following the introduction of rainbow trout in the 1930s. In lakes, competition from lake trout *Salvelinus namaycush* stocked in Lake Tahoe (Nevada) before the turn of the century and from kokanee *Oncorhynchus nerka* introduced into Independence Lake (California) had substantial impact on the Lahontan cutthroat trout (Gerstung 1988). The presence of kokanee also affects westslope cutthroat trout in lakes in Glacier National Park, Montana, because the two species compete for plankton (Marnell 1988).

Another indication of the inherent lack of competitive ability of the cutthroat trout is the fate of hundreds of millions of Yellowstone cutthroat eggs and fry that were shipped to waters around the world from Yellowstone National Park. Although these fish were stocked in several dozen states in the USA and in several Canadian provinces and in several other countries, very few viable populations were established in the presence of competitive species (Varley and Gresswell 1988).

A limited evaluation of interactions between introduced nongame species and cutthroat trout has been conducted. Although redside shiners and longnose suckers coevolved with westslope cutthroat trout, they are not native to Yellowstone Lake. Following introduction of the longnose sucker in the 1920s and the redside shiner in the 1950s into Yellowstone Lake, research indicated no evidence of direct competition with Yellowstone cutthroat trout, and it was suggested that both species filled previously vacant niches (Biesinger 1961; Gresswell and Varley 1988). However, in a Colorado lake, competition for *Daphnia* was noted between the native longnose sucker and an introduced population of cutthroat trout derived from the greenback cutthroat trout (Trojnar and Behnke 1974).

Alternatives to Competitive Exclusion

Some evidence suggests that, for some cutthroat trout populations under certain circumstances, interaction with introduced salmonids may not eliminate the cutthroat trout. Coastal cutthroat trout appear capable of maintaining their integrity in the presence of nonnative salmonids (Nicholas 1978). Westslope cutthroat trout also show some ability to compete. For example, in the upper Flathead River drainage, Montana, they have been very resistant to introductions of rainbow trout (Liknes and Graham 1988, this volume). In the St. Joe River, Idaho, Petrosky and Bjornn (1985) observed little short-term detrimental effect on westlope cutthroat trout resulting from the stocking hatchery rainbow trout. The numbers of cutthroat trout remained stable following stocking except in one trial, where the stocking level of 3.5 fish per linear meter of stream was higher than that normally utilized by management agencies.

For Lahontan cutthroat trout, the Humboldt River race is thought to be more resistant to introduced salmonids than Lahontan cutthroat trout in other basins (Gerstung 1988). Reasons for this are not known, but Behnke (1979) suggested that development of the race under the particularly severe climatic conditions it experiences has produced an adaptable genotype that is capable of resisting replacement by nonnative trouts.

In a habitat that is optimal for them, cutthroat trout might be expected to be most capable of holding their own against introduced competitors. By closely observing competition between cutthroat trout and introduced salmonids, we may learn what that optimal cutthroat trout habitat is. For westslope and, possibly, Yellowstone cutthroat trout, it may consist of the high-gradient portions of the heads of small streams. In an extensive survey of the fish community of the Salmon River drainage in Idaho, Platts (1974) found that westslope cutthroat trout density peaked at a channel gradient of about 10%, which was higher than that for peak densities of bull, rainbow, or brook trouts. However, it is possible that those cutthroat trout represent remnant populations that exist there not because it is optimal habitat for them, but because it is less optimal for other salmonids. High stream gradient appears to have a negative influence on abundance of brown trout (Hermansen and Krog 1984), rainbow trout (Kennedy and Strange 1982), and brook trout (Chisholm and Hubert 1986). Controlled experiments designed to identify optimal habitat for cutthroat trout subspecies would provide valuable insights.

Replacement, an alternative to competition as a mechanism causing cutthroat trout decline, is the concept whereby cutthroat trout are eliminated by factors such as habitat degradation or harvest by anglers and then an introduced species fills the void. Cutthroat trout are very vulnerable to angling (Varley and Gresswell 1988) and are more easily caught than are brook trout (MacPhee 1966) or other salmonids. Replacement, at least in stream environments, may be an irreversible process. As observed by Moyle and Vondracek (1985) in Martis Creek, California, once a cutthroat trout is replaced by a member of another salmonid species, it is unlikely that space will ever be regained by a cutthroat trout.

In some situations where rapid species shifts occur, the replacement process appears to be a more plausible explanation than interspecific competition. Further study is needed to more clearly define the processes of competition and replacement, as they affect cutthroat trout subspecies, and the rates at which they operate.

Acknowledgments

Comments by Bob Gresswell, Kurt Fausch, and two anonymous reviewers greatly improved the manuscript. Funds for field work were provided by a Faculty Research Grant from Idaho State University.

References

Andrusak, J., and T. G. Northcote. 1971. Segregation between adult cutthroat trout (*Salmo clarki*) and

Dolly Varden (*Salvelinus malma*) in small coastal British Columbia lakes. Journal of the Fisheries Research Board of Canada 28:1259–1268.

Behnke, R. J. 1979. The native trouts of the genus *Salmo* of western North America. Report to U.S. Fish and Wildlife Service, Denver, Colorado.

Behnke, R. J. 1988. Phylogeny and classification of cutthroat trout. American Fisheries Society Symposium 4:1–7.

Biesinger, K. E. 1961. Studies on the relationship of the redside shiner (*Richardsonius balteatus*) and the longnose sucker (*Catostomus catostomus*) to the cutthroat trout (*Salmo clarki*) population in Yellowstone Lake. Master's thesis. Utah State University, Logan.

Birch, L. C. 1957. The meanings of competition. American Naturalist 91:5–18.

Bisson, P. A., and J. R. Sedell. 1984. Salmonid populations in streams in clearcut versus old-growth forests of western Washington. Pages 121–129 *in* W. R. Meehan, T. R. Merrell Jr., and T. A. Hanley, editors. Fish and wildlife relationships in old-growth forests. American Institute of Fishery Research Biologists. (Available from J. W. Reintjes, Route 4, Box 85, Morehead City, North Carolina.)

Bisson, P. A., K. Sullivan, and J. L. Nielson. In press. Channel hydraulics, habitat utilization, and body form of coho salmon, steelhead trout, and cutthroat trout in streams. Canadian Journal of Fisheries and Aquatic Sciences.

Chapman, D. W. 1966. Food and space as regulators of salmonid populations in streams. American Naturalist 100:345–357.

Chisholm, I. M., and W. A. Hubert. 1986. Influence of stream gradient on standing stock of brook trout in the Snowy Range, Wyoming. Northwest Science 60:137–139.

Cummings, T. R. 1987. Brook trout competition with greenback cutthroat in Hidden Valley Creek, Colorado. Master's thesis. Colorado State University, Fort Collins.

Fausch, K. D., and R. J. White. 1981. Competition between brook trout and brown trout for positions in a Michigan stream. Canadian Journal of Fisheries and Aquatic Sciences 38:1220–1227.

Gerstung, E. R. 1988. Status, life history, and management of the Lahontan cutthroat trout. American Fisheries Society Symposium 4:93–106.

Gibson, R. J. 1973. Interactions of juvenile Atlantic salmon and brook trout. International Atlantic Salmon Foundation Special Publication 4:181–202.

Glova, G. J. 1984. Management implications of the distribution and diet of sympatric populations of juvenile coho salmon and coastal cutthroat trout in small streams in British Columbia, Canada. Progressive Fish-Culturist 46:269–277.

Glova, G. J. 1986. Interaction for food and space between experimental populations of juvenile coho salmon (*Oncorhynchus kisutch*) and coastal cutthroat trout (*Salmo clarki clarki*) in a laboratory stream. Hydrobiologia 131:155–168.

Gresswell, R. E., and J. D. Varley. 1988. Effects of a century of human influence on the cutthroat trout of Yellowstone Lake. American Fisheries Society Symposium 4:45–52.

Griffith, J. S. 1972. Comparative behavior and habitat utilization of brook trout (*Salvelinus fontinalis*) and cutthroat trout (*Salmo clarki*) in small streams in northern Idaho. Journal of the Fisheries Research Board of Canada 29:265–273.

Griffith, J. S. 1974. Utilization of invertebrate drift by brook trout (*Salvelinus fontinalis*) and cutthroat trout (*Salmo clarki*) in small streams in Idaho. Transactions of the American Fisheries Society 103:440–447.

Hanson, D. L. 1977. Habitat selection and spatial interaction in allopatric and sympatric populations of cutthroat and steelhead trout. Doctoral dissertation. University of Idaho, Moscow.

Hartman, G. F., and C. A. Gill. 1968. Distributions of juvenile steelhead and cutthroat trout within streams in southwestern British Columbia. Journal of the Fisheries Research Board of Canada 25:33–48.

Hearn, W. E. 1987. Interspecific competition and habitat segregation among stream-dwelling trout and salmon: a review. Fisheries (Bethesda) 12(5):24–31.

Hermansen, H., and C. Krog. 1984. Influence of physical factors on density of stocked brown trout in a Danish lowland stream. Fisheries Management 15: 107–115.

Hickman, T. J., and D. A. Duff. 1978. Current status of cutthroat trout subspecies in the western Bonneville Basin. Great Basin Naturalist 38:193–202.

Johnson, S. W., J. Heifetz, and K. V. Koski. 1986. Effects of logging on the abundance and seasonal distribution of juvenile steelhead in some southeastern Alaska streams. North American Journal of Fisheries Management 6:532–537.

Jones, D. E. 1978. A study of cutthroat and steelhead in Alaska. Alaska Department of Fish and Game, Annual Performance Report, AFS 42-6, volume 19, Juneau.

Kennedy, G. J. A., and C. D. Strange. 1982. The distribution of salmonids in upland streams in relation to depth and gradient. Journal of Fish Biology 20:579–591.

La Rivers, I. 1962. Fishes and fisheries of Nevada. Nevada State Printing Office, Carson City.

Liknes, G. A., and P. J. Graham. 1988. Westslope cutthroat trout in Montana: life history, status, and management. American Fisheries Society Symposium 4:53–60.

MacPhee, C. 1966. Influence of differential angling mortality and stream gradient on fish abundance in a trout–sculpin biotope. Transactions of the American Fisheries Society 95:381–387.

Marnell, L. F. 1988. Status of the westslope cutthroat trout in Glacier National Park, Montana. American Fisheries Society Symposium 4:61–70.

Moffitt, C. M., and T. C. Bjornn. 1984. Fish abundance upstream from Dworshak Dam following exclusion of steelhead trout. Idaho Water and Energy Re-

sources Research Institute, Project WRIP/371404, Technical Completion Report, Moscow.
Moyle, P. B., and B. Vondracek. 1985. Persistence and structure of the fish assemblage in a small California stream. Ecology 66:1–13.
Nicholas, J. W. 1978. Life history differences between sympatric populations of rainbow and cutthroat trouts in relation to fisheries management strategy. Pages 181–188 *in* J. R. Moring, editor. Proceedings of the wild trout–catchable trout symposium. Oregon Department of Fish and Wildlife, Corvallis.
Nilsson, N. A., and T. G. Northcote. 1981. Rainbow trout (*Salmo gairdneri*) and cutthroat trout (*S. clarki*) interactions in coastal British Columbia lakes. Canadian Journal of Fisheries and Aquatic Sciences 38:1228–1246.
Petrosky, C. E., and T. C. Bjornn. 1985. Competition from catchables—a second look. Pages 63–68 *in* F. Richardson and R. H. Hamre, editors. Wild trout III. Federation of Fly Fishers and Trout Unlimited, West Yellowstone, Montana.
Platts, W. S. 1974. Geomorphic and aquatic conditions influencing salmonids and stream classification, with application to ecosystem classification. U.S. Forest Service, Billings, Montana.
Platts, W. S., and R. L. Nelson. In press. Fluctuations in trout populations and their implications in land-use evaluation. North American Journal of Fisheries Management 8.
Pratt, K. L. 1984. Habitat use and species interactions of juvenile cutthroat, *Salmo clarki lewisi*, and bull trout, *Salvelinus confluentus*, in the upper Flathead River basin. Master's thesis. University of Idaho, Moscow.
Trojnar, J. R., and R. J. Behnke. 1974. Management implications of ecological segregation between two introduced populations of cutthroat trout in a small Colorado lake. Transactions of the American Fisheries Society 103:423–430.
Varley, J. D., and R. E. Gresswell. 1988. Status, ecology, and management of the Yellowstone cutthroat trout. American Fisheries Society Symposium 4: 13–24.
Varley, J. D., and P. Schullery. 1983. Freshwater wilderness: Yellowstone fishes and their world. Yellowstone Library and Museum Association, Yellowstone National Park, Wyoming.
Wiens, J. A., and J. T. Rotenberry. 1981. Habitat associations and community structure of birds in shrub steppe environments. Ecological Monographs 51: 21–42.